T0184684

Paradoxical Urbanism

Malcolm Miles

Paradoxical Urbanism

Anti-Urban Currents in Modern Urbanism

Malcolm Miles
Bradford-on-Avon, UK

ISBN 978-981-15-6343-0   ISBN 978-981-15-6341-6   (eBook)
https://doi.org/10.1007/978-981-15-6341-6

Cover Pattern © Melisa Hasan

This Palgrave Pivot imprint is published by the registered company Springer Nature
Singapore Pte Ltd.
The registered company address is: 152 Beach Road, #21-01/04 Gateway East, Singapore
189721, Singapore

INTRODUCTION

I wrote this book from discontent with urban redevelopment as it currently occurs, which is in service to global capital, against the interests of social justice and political accountability. It is an obvious argument but I also wanted to remember that cities might be where human potential is most likely to be realised, in an ambience of mobility, among people of different backgrounds, which produces an enhanced sense of self and a capacity for agency. But this is precisely what most redevelopment schemes deny, constructing compounds for elites and empty investment opportunities. Each new centre turns adjoining areas into margins, and the underlying concept of the City is manipulated as a scene of waterfront vistas and steel-and-glass corporate towers. Again, this is familiar, and well covered in the critical literature of urbanism (beside critiques of the notion of a creative city as a defence of gentrification). All this represents a policy failure, and a political malaise. Meanwhile, with the hollowness of representational politics, there is a rise in non-violent direct action, which has a growing literature. My aim was not to add to that, nor to the literature of urban redevelopment; it was instead to ask if there are underlying flaws *within* the idea of the modern city which limit its potential for well-being, and need to be addressed. This is historical, and questions the basis of how cities are imagined and designed.

I identify two such flaws: the enchantment of a rural idyll, especially in English culture since the sixteenth century; and an attraction to self-contained systems in European thought from the seventeenth century.

The first begins with an aristocratic dream of timeless bliss as reaction to the intrigues of the court and the onset of modernisation. Its price includes land enclosures, the dislocation of villagers, and violent enforcement. The second is a revolution in thinking in René Descartes' *Discourse* of 1637, which expresses an almost desperate search for certainty, looking to un-changing systems such as geometry and mathematics, written during the Thirty Years' War.

Taking these two critiques, I argue that the potential for city living—being among people with different dreams, being able to act as a citizen—is limited by a fear of conflict which leads to compartmentalisation; hence the projection of a rural idyll, the spread of suburbia, and in another way, the dehumanising functionalism of modernist planning and design. This is complex, and this short book is a modest contribution to its understanding. But I hope the connection between the rural idyll and the city-as-machine will provoke discussion. I hope, too, that readers will ask themselves what else is possible.

I focus on Europe because, in a short book, I cannot adequately offer worldwide scope; and because I write from my own experience, when possible about places I can visit. It is more than ten years since I took a long-haul flight (for environmental reasons). Besides, I regard myself as a European.

Throughout the book, I try to explain rather than mask complexities, but also allow myself nuances. I enjoyed a day in the landscaped park at Rousham, and I happen to like Descartes' writing, for instances, but am critical of what they mean. Perhaps ambivalence is a writer's occupational hazard, while critical insights emerge on an axis of creative tension between the polarities which define an argument.

In the end—except it is never an end because effects become causes and causes similarly become effects—it is the means which count. These are often local and ephemeral; yet, facing modernist urbanism's ruinous instrumentalism, they begin to create an alternative urban imaginary.

CONTENTS

LIST OF FIGURES

CHAPTER 1

Colliding Utopias

Abstract This chapter argues that the concept of the City—distinct from specific cities—is idealised, in part in reaction to industrialisation and cultural representations of inner-city dirt and crime. This negativity denies the benefits of freedom from rural ties to the land and kin, favouring nostalgia for imagined pasts. Yet the conditions of city life – living among people from different backgrounds, social and geographical mobility, and the agency of citizens—are liberating. The chapter reflects on the meanings of the terms 'city' and 'urban'; and on the contradictory term urban village. It observes the rise of suburbia, and the search for stability in refusals of the contingencies of city living. Finally, it turns to sociology, and a gap between a regressive nostalgia and a recognition of engagement in city life's perpetual changes.

Keywords City • Urban • Urban village • Community • Society • Suburb

Words carry baggage. Their meanings are mediated between the rules of a language and its everyday uses within the social, cultural, economic and political conditions of a society. One aim of this chapter is to draw out different meanings and associations which indicate how the concepts 'city' and 'urban' are understood, and might be questioned. Another is to suggest that certain conditions found in cities, such as a high density of population, or a diversity of people and interests, may be either threatening or a means to a liberating way of life which is not possible in smaller settlements. In the next chapter I examine the allure of village life, as mythicised in the nineteenth and twentieth centuries, arguing that it undermines the potential for liberation in cities. Here, I begin by looking at words, then move to conditions and the range of representations which shaped how conditions are apprehended.

WORDS AND PLACES

The words city and urban allude to large human settlements, but with different associations. In English, urban is rarely used before the nineteenth century,[1] but urbane, from the same root, means courteous or educated, linked to living in a city with institutions of learning and an elite able to develop ideas of taste. This may seem to be of interest to philologists but not much help in ordinary life yet words and meanings exhibit frames of meaning which are ingrained but not fixed. The idealism attached to the concept of the city derives in part from the Greek word *polis*, which means the democratic city, informing the English word political. But idealism reveals its undercurrent here, too, because the policy of classical Athens was the preserve of around 5% of its population, all wealthy, free-born men. The political is a matter of power relations—who is included or not in determining the future, and the image, of a city—while the interests of different groups may be not only oppositional, but also overlapping. As political scientist Margaret Kohn argues,

> Urban space is often characterised by contrasts and contradictions that exist side by side; rich and poor; production, consumption and reproduction; decay, renewal and reappropriation. Their juxtaposition in space is a glaringly visible reminder of the impossibility of achieving complete control, order and homogeneity. It is the quality that makes urban space at once threatening, fascinating and haunting.[2]

Kohn cites critical theorist Seyla Benhabib's discussion of the term actuality to denote, not a situation as it is but, 'what it could be but is not.'[3] In 1960s Paris, the Situationist group of students and artists made unplanned detours in the city to find sites of revolutionary pasts, and, 'a poetry made by the communal appropriation of the past in the present.'[4] Today, in not dissimilar detours, urban explorers transgress spatial and legal boundaries to find what might be equivalent traces in derelict interiors, underground passages and de-industrialised sites. For geographer Oli Mould they, 'defy the prevailing narrative' of late capitalism to see signs of alternative pasts and futures.[5] I return to this in Chap. 5.

To return to the terms urban and city, sociologist Richard Sennett writes of three kinds of space in medieval Paris: the *cité* included the royal palace and Cathedral; merchants lived and traded in *bourgs* on both banks of the Seine; everyone else lived in *communes*, with no defensive walls. Outside the *cité*, buildings were separated only by the space necessary for circulation. The pathways of one commune did not link to those of another. Shop-keepers used hired thugs to attack their rivals' customers. Street life was vicious.[6]

Sennett's realist picture of medieval city life contrasts with the idealised picture of cities in, say, late medieval and early Renaissance art: white towers and crenelated walls adorned by brightly coloured pennants; and stone gates to protect the city from the world outside. But the prevailing idea of a city—or *the* city, like a citadel—is shaped by such representations, a legacy of the culture of a privileged class which retains currency as a vehicle for projection of how cities might be, of what would be nice.

For geographer Edward Soja, city plans are determined by, 'mental or cognitive mappings of urban reality and the interpretive grids through which we think about, experience, evaluate and decide to act.'[7] Interpretive grids become ingrained but the role of academic work is to interrupt this, rupturing any generalised notion of what a city is in favour of investigation of the specific conditions and experiences of what cities (plural) are. For Soja, this questions a convention that villages came first, and grew into towns, then cities. He shows that this was not always the case by citing archaeological evidence from Çatal Hüyük, Anatolia.[8]

In Çatal Hüyük, communication was by rooftops reached by ladders from courtyards. As its population grew the city expanded by building outward. Exterior walls had no windows or doors, and might be reinforced. But this *ad hoc* expansion was not informal but orderly, as archaeologist James Mellaart writes, using standardised house plans, brick sizes, doorway heights, and measures of a human hand (8 cm) and foot (32 cm).[9] As cultivation and herding spread in a widening belt around the city, shepherds built huts for overnight stays, which were the first villages.

Çatal Hüyük is unlike a modern Western city in both form and trajectory but contributes to an understanding of cities as sites of high-density dwelling, division of labour, and a specific spatial practice. The rest is a matter of negotiation, which is also what cities are about.

REPRESENTATIONS AND NARRATIVES

In sharp contrast to Çatal Hüyük, classical Athens is said to express the democratic ideal. As noted above, this is illusory. Athenian society relied on slave labour and foreign conquest; it had a limited democracy which excluded women, foreigners and slaves, leaving 5% of the population, all free-born, wealthy men, to participate in the assembly (*pnyx*), where political decisions were made. In the *agora*, a marketplace surrounded by booths for administrative purposes, a colonnade where citizens of sufficient leisure could converse offered debate. Yet there were few public events in the *agora* and the myth of Athenian democracy persists because that is how this history was written, by a class for whom the presiding image was convenient.

Other prevailing images are informed by medieval sites, such as Sienna, or San Gimignano in Tuscany. Geographer John Gold notes that Cumbernauld in Scotland, built in the 1960s as a series of concrete structures on a ridge, has, 'the symbolic appearance of a medieval Italian hilltop citadel.'[10] Life in a medieval citadel was, as in Paris, more violent than that expected in Cumbernauld; and Gold makes the point critically, not to affirm the idea. But myths of the city are persistent. Industrial cities conjure smoke and grime; but Sennett writes that, in the *bourgs* of Paris, 'the economy promised to set [citizens] free from the inherited dependence embodied in the feudal labour contract.'[11] For an emerging bourgeois class a *bourg* was free from feudal ties. Merchants formed trade networks, such as the Hanseatic League of ports around the Baltic Sea, and negotiated terms with ruling dynasties. On their gates, these ports displayed the message, 'city air makes one free' (*Stadt Luft macht frei*).[12] As Sennett writes, 'Profit lay on the horizon of the possible, in the land of perhaps, toward which one travelled as often as one could, from which one often failed to return.'[13]

A transition from the medieval and Renaissance to the modern city occurs in the eighteenth and nineteenth centuries, with industrialisation and rapid urban expansion. In England, the process began in the seventeenth century when an emerging class of merchant-investors in London established production sites across the country, taking risks and gaining rewards (or losses). Such changes in social and economic patterns mould changes in a prevailing image of the city, a signifier floating above the material actualities of cities. For geographer James Donald the questions is whether, 'If the city is an imagined environment, and modernity is an attitude more than it is an epoch,' dominant images and metaphors which determine how the modern city is mediated can be identified.[14]

In nineteenth-century England, the rise of industrialisation—from machines in agriculture which produced evictions of rural populations, to factories in the towns and cities to which those populations migrated—the urban image was dark. The negativity is present today in an alignment of inner-city areas to crime, and the privileged status and high property values of non-industrial districts. Of course, negativity reflects real abjection; but it is also an image which colours future development. Donald cites two kinds of evidence for urban negativity: Friedrich Engels' account of housing in Manchester in the 1840s,[15] and photography, as in Jacob Riis' pictures of New York, and Eugène Atget's of Paris.

Social research and culture intersect in Charles Dickens' novels. He sets part of the action of *Oliver Twist* (1837) in the high-density housing, called rookeries, of central London, affirming that such sites breed crime but inserting a humanity otherwise overlooked; he writes of the marginal life of the Thames mudflats, where people live by fishing corpses from the river, in *Our Mutual Friend* (1864); and represents a northern industrial city in *Hard Times* (1854) as a smoke-drenched wasteland which is, not unconnectedly, a location of mechanistic learning—facts not feelings—in schoolmaster Mr. Gradgrind's regime. Later, in the work of Arthur Conan Doyle and Joseph Conrad, villains inhabit dark, narrow, twisting alleys shrouded in fog, plotting crimes, betrayals, or the end of civilisation.[16] The dark alleys become aligned to a primitive past, as if a modern city should open them to light and civic order. The fog is not primordial, however, but the real by-product of fossil-fuelled industry.

Reports arising from social research and a Royal Commission on housing in 1884–85 led to liberal reformism: the improvement of material conditions for the poor as a way to increase productivity and prevent insurrection. The rest is fiction. But it lingers in public imaginations while literary representations outlast the conditions they reflect. For social historian Oswald Spengler in the 1920s, a generalised, received negativity becomes universal decline attributable to the city. Literary historian Richard Lehan summarises Spengler's *Decline of the West* (1926) as an organic theory of society:

> ... each culture follows a biological pattern of growth and decline, vitality (stemming from the land) and decadence (spreading from the city). As one moves away from the natural rhythms of the land, instinct is replaced by reason, nature and myth by scientific theory, and a natural marketplace (barter and exchange) by abstract theories of money.[17]

Spengler, then, constructs a meta-narrative of entropy, in which the city is civilisation's fall, itself the protagonist, rather than Conrad's spies and criminals. For Spengler, 'the whole life blood of broad regions is collected [in cities] while the rest dries up.'[18] Spengler constructs a city–province dualism, and presents this as a universal, inevitable, a-historical narrative from which there is no exit: from an enchanted land to wasteland, then darkness. Something not unlike this narrative, but in a more complicated form, pervades T. S. Eliot's epic poem *The Waste Land* (1922).

Urban planners do not make decisions based on their readings of Spengler or Eliot, yet an a-historical narrative may appeal in times of discontinuity. For example, the Chicago school of urban sociologists and planners in the 1920s and 1930s saw the city—based on Chicago with a central business district, transitional zones for migrants, and rings of suburbs of increasing affluence—as moulded by a logic as inevitable as Spengler's; and used a biological metaphor of plant growth for urban change. This universalised and de-politicised the city on the model of Chicago because, it seemed, politics was disagreement between factions, while expert or technocratic knowledge could resolve conflict, as if neutrally. So, if Spengler and Eliot, and the Chicago urbanists, inscribe a neutral model from fear of social breakdown, they do this to undermine contesting claims made by diverse publics. Another product of fear of conflict is the so-called urban village.

The Urban Village

The words village and urban have opposite associations, one with an imagined countryside and the other with everyday city life. Developers use the label urban village to market inner-city redevelopment schemes. Usually these consist of housing compounds for new elites but Paddington Urban Village, under construction in Liverpool in 2020, is a knowledge quarter—quarters are also popular with developers, lending a feeling of an old city—between the city centre and one of its universities. The appeal of the label urban village rests on the myth of lost contentment, its currency deriving from an idea of community. A hoarding outside the demolition site which was once the Heygate Estate in Southwark, London, proclaims:

> Be part of it
> The Life
> The Heart
> The Elephant

These ten words assert, in order, community; humanity; and ... Elephant, the name of the redevelopment scheme, as if not requiring explanation. (Fig. 1.1) It replaced social housing for three thousand people in medium-rise blocks built in 1974, with large rooms, near central London. Designed by local authority architect Timothy Tinker, the estate had another thirty years of viability before major refurbishment, but the local authority ran it down and leased it for dystopian film sets. The developer's undertaking to build affordable housing on-site was broken.[19] Journalist Peter Walker concludes, 'poor residents are banished from central London ... replaced by wealthy incomers and buy-to-let investors.'[20] Yet Tinker reflected, 'I don't think it was a failed estate ... what we provided in concrete and brick was relatively OK.'[21] The Elephant is not an urban village, consisting of tower blocks, but still trades on the rhetoric of community, its regressive image being the successor to the suburban ideal of the 1880s–1950s.

Fig. 1.1 Be part of it: Redevelopment site hoarding, Elephant, London. (Author's photograph)

Metropolitan transport and mortgage finance enabled the move to sub-urbia,[22] but a myth of security and continuity produced the desire, open only to an aspirant working-class and lower-middle class. Sociologist Elizabeth Wilson notes, 'for the poorest, the suburban ideal remained beyond their reach.'[23] Inner-city districts became repositories for social immobility although also, under the radar, offering a real sense of belonging through the practical care of neighbours, credit in corner shops, and extension of domestic space into the public–domestic space of the street.[24] Suburbs replaced that real community, accessed only in tacit understandings, with a notion of community fuelled by nostalgia for an imagined, unreal rural idyll (recreated in polite form in back gardens).

Fear of the city redirects attention to pasts which are remote enough to accept almost any projection, as if what is absent can be reclaimed in a fabled elsewhere. This illusion is brittle, and breeds a fear of outsiders whose interruption might shatter its purity. In the 1980s, an ideal of a conflict-free city, laced with North American nostalgia denoted by the white picket fences was built in the Disney township of Celebration, Florida.[25] With colonial-style houses, Celebration has a Town Hall, but is run by a Disney employee, a Town Manager. The appeal of Celebration, and the New Urbanism it instantiates, rests on a fear of conflict exacerbated by negative representations of the city in literature, social research and popular culture.

It is also gendered: men, being brave heroes who defy the odds, engage with city life but women are confined to domestic roles and spaces, and conformity. Wilson cites a magazine, *Good Housekeeping*, in the 1920s, '"The busy men leave on early trains, and are at once plunged into the rush of their accustomed life" ... but the young wife is left behind.'[26] Men plunge and rush into voluntary ties; women dwell in routines except on trips to department stores. Men, however, loan then new appliances to relieve drudgery. Wilson quotes planner Patrick Geddes in 1915: 'we may begin to emancipate Cinderella, no longer depress her through slavery into charwoman or crone.'[27] She continues that in 1950s Britain, women in suburbs and new towns were allowed to have taste in decorating and furnishing their new homes; but planners and designers had no time for three-piece suites, wallpaper, or net curtains; nor for DIY interventions which messed up open-plan living, eating and kitchen areas.[28] Modernism was progressive, but as a top-down style.

Moving to the 1960s in North America, Sennett reads white, North American suburbia as reproducing a fantasy of community, enforced by conformity and an imperative to belong, which assumes a new violence. He writes that when fear of the unknown takes over, 'the acceptable future can be conceived only in the same form as the present.'[29] Planning academic Leonie Sandercock summarises this as the maintenance of falsehood:

> [Sennett] identifies a "myth of a purified community" which is bred out of the way human beings learn … how to lie to themselves in order to avoid new experiences that might force them to endure the pain of perceiving the unexpected, the new, the otherness around them. … Sennett then makes a fascinating leap to the planning profession … arguing that … city building has proceeded out of this very same craving for ordering, for creating a unified and comprehensible whole from the disparate and messy parts, for eliminating uncertainty and disorder.[30]

Sennett characterises suburban life as a perpetual adolescence, never reaching the mature self which—Sennett draws on philosopher Hannah Arendt—is produced in the perceptions of others. Hence suburbanites reject outsiders while appealing to family values:

> This society of fear, this society willing to be dull and sterile in order that it not be confused or overwhelmed, thus shares something with the first middle-class families of the industrial city … and the family becomes a place of refuge in which the parents try to shield their children, and themselves, from the city.[31]

Sanderock extends Sennett's argument, however, citing Celebration:

> For increasing numbers of homeseekers in the United States this is the new Utopia – an escape from what are seen as the ugly realities of urban life into the civic-minded communities of front porches and safe, shady streets, and a return to 1950s-style family values, a time of innocence in which, according to the promotional video, the biggest decision is whether to play kick the can or king on the hill.[32]

A Utopia stands as counterfoil to urban dis-ease, but is illusory, avoiding the negotiation and acceptance of living in a contingent realm which is potentially liberating. A purified nowhere with its attendant fictional notion of belonging is thus destructive.

Unpacking the construct of an urban village is one way to interrupt the rhetoric. The urban village carries the legacy of suburban security and its implicit denials, now redesigned for a class of aspirant young professionals. The label urban village is also used to market nodes of consumption. Both pertain to gentrification. An urban village in Manchester is promoted in the local press as where to find, 'wine and cocktail bars, restaurants, shops, a florist and even a barber all bunkered beneath the flyover [the Mancunian Way].'[33] Eleven new shops occupy old shipping containers, including a vintage clothing store. The developer is quoted, 'From becoming this food and drink destination ... we're hoping it will evolve into an urban village.'[34] Hmmm ...

Perhaps an alternative model is glimpsed in a project seeking something not unlike a village ambience, but decidedly urban, the Byker Wall, Newcastle, designed by Ralph Erskine in the 1960s. Byker Wall replaced a demolished area of nineteenth-century working-class housing, with twelve hundred flats in tower blocks of varying height, with balconies and minor design variations. The towers are clustered round a garden courtyard with play areas, and shield residents from the noise of an adjacent fast road. A comparison could be made with Karl Marx Hof in Vienna, built in the 1920s (Chap. 6). The courtyard is visible from flats above, making a viable space for shared watching-over of children. Refurbished in 2014, the site is recognised by UNESCO. It is clearly modernist in its design and use of materials, and has no need for rural nostalgia while maintaining a sense of belonging integrated with the rest of the city.

Nostalgia

So far I have written colloquially, quoting academic sources but in course of a story rather than a narrative of urbanism. I move now to more academic terrains, and the presence of nostalgia in attitudes to cities. In the 1960s, Jane Jacobs campaigned to preserve districts of Boston and New York threatened by clearance. This was progressive, opposing development serving property rather than community interests. But it was regressive in assuming that a community is a coherent body of public opinion; and in assuming that an informal mixing of strangers occurred spontaneously in urban streets. Jacobs worked especially in Greenwich Village, New York, a neighbourhood with a bohemian past, associated with writers including Gamel Woolsey and Djuna Barnes. Greenwich Village had (and has) a range of small shops and bars, and a human scale, low-rise streetscape of residential and small-scale commercial uses. In part thanks to Jacobs and a preponderance of opinion formers among its residents, Greenwich Village did not undergo the aggressive gentrification of SoHo (the old garment district, where artists' lofts gave way to design practices, then niche-market outlets, as rent levels increased). Working with local residents, Jacobs drew attention to the desirability of Greenwich Village as it was. But when she observed the interactions of people in the streets of Greenwich Village, she did not see a typical inner-city environment. As journalist Roberta Gratz remarks,

> Jacobs saw the city as a holistic organism at a time when prevailing movements insisted on breaking it into divisible parts. Recognising that cities have been and will always be under some kind of stress, never reaching a clear equilibrium, she illustrated that, like natural organisms, cities develop their own form of health if not inappropriately interfered with.[35]

Inappropriate interference was associated with major road building schemes, and a vision of a city seen through a car windscreen. Jacobs noted that redevelopment forced out tenants in low- and mid-cost housing, and that the design of commercial buildings asserted a form of, 'architectural wilfulness.'[36] Her campaigns for preservation (adaptation not demolition) and mixed-use spaces have had an important legacy, and I do not question it. My question is whether this still rested on a notion of community, and the commons (of informal mixing in shared spaces, as equals) which was unreal.

Sennett writes of the charms of Greenwich Village, where he, like Jacobs, lives; he mentions that resident Italian families mix with tourists, and apartment blocks, 'contain elderly people who have guarded their cheap housing and live intermixed with newcomers who are richer and younger.'[37] He remarks that the writers and artists who remain are, like himself, ageing, bourgeois bohemians for whom the Village provides sympathetic space. But Sennett accepts the limits of his picture:

> Diversity in the Village works this way: ours is a purely visual agora. There is nowhere to discuss the stimulations of the eye … no place they can be collectively shaped into a civic narrative, nor, perhaps more consequentially, a sanctuary which takes account of the disease-ravaged scenes of the East Village. Of course the Village … offers myriad formal occasions in which our citizens voice civic complaint, outrage. But the political occasions do not translate into everyday social practice on the streets; they do little, moreover, to compound the multiple cultures of the city into common purposes.[38]

Jacobs was less critical, seeing the street as producing a public sphere. I do not argue against a need for a public sphere but question whether there has been one, ever, where people of all classes exchange opinions on the basis of equality.

My worry, then, is that notions of a lost public sphere are as nostalgic as notions of village community; and whether this embellishes the fear of difference found in suburbia. Architect Gwendolyn Wright sees Greenwich Village, Morningside Heights, and Harlem, all distinctive New York neighbourhoods, as where creative groups have migrated, such as the Beat poets and Abstract Expressionist painters in Greenwich Village; each has an, 'emphatic localism ... self-consciously separate from the rest of the city, including the other nodes of creativity ... [so that] outsiders were pariahs in each enclave.'[39] Enclaves, she argues, can be porous; but the critical realism which informs her remarks is absent in Jacobs' legacy. That is my worry about Jacobs' otherwise laudable efforts, and the urban village scam which draws in part on her legacy. For example, landscape architect Peter Neal writes in support of urban villages,

> Historically, the form and purpose of the village provide important precursors to the majority of our towns and cities. Many of the innate qualities found in villages have been retained in perpetuity to provide a type of metropolitan code ... Throughout time we have seen human life swing between two poles, from movement to settlement. From the earliest settlements a sense of community was forged to meet the needs of nutrition and reproduction in relative safety.[40]

This makes a number of generalisations. A sense of quasi-rural community is assumed as a universalised mechanism of development which ensures safety, against the threats of inner-city disorder. Neal concludes (writing of North American cities) that, 'whatever remains of authentic urbanism can be the foundation for regeneration, once again serving as centres of innovation, strength and enduring charter.'[41] Without wishing to labour the point, this rests on a supposed continuity; on a non-situated notion of authenticity, and, naïvely on the good intentions of the development industry. I leave it there.

COMMUNITY AND SOCIETY IN EARLY SOCIOLOGY

The nostalgia which haunts urbanism follows a sense of rural nostalgia more carefully set out in early sociology, in the 1880s to 1900s. But while sociology affirms the nostalgia for village continuity and safety it also recognises prospects for engagement, particularly in big (or metropolitan) cities, which offer an exit from the bind. I turn now to the work of two German sociologists, and the emergence of such a trajectory.

Living in rural Germany, Ferdinand Tönnies identified a sense of community in village life (*Gemeinschaft*), in contrast to the city's shifting social interactions (*Gesellschaft*) reflecting individualism and transactional relations.[42] Although community is felt by neighbours in a city, the abstraction of money reinforces their individual isolation. The modern city dweller is lonely, and surrounded by a constantly changing situation. Sociologist David Frisby cites Tönnies saying that, 'the social theorist is presented with the distinctive problem of ... capturing the fleeting and the transitory.'[43] Transience produces a desire for a lost permanence, or for community as a trans-historical, trans-national state of mind, like Nature.

This theory is presented in Tönnies' major work, *Gemeinschaft und Gesellschaft* (1887). Its first section defines community and society as, respectively, structures of household, clan and ethnic group as well as medieval guilds; and of the market economy and commodity exchange. Village community expresses a common will, but society is moulded instead by contracts and wage labour, and a strategic will, looking instrumentally to future advantage in transactions. Tönnies presents these conditions objectively, like a historian, and sees society emerging to replace community through the division of labour and capital accumulation of modern cities. He uses the term Communism for village mutuality, if in a de-politicised way. The book's context, still, is quite political.

Gemeinschaft und Gesellschaft was written after the unification of Germany and its defeat of France in 1870 (gaining two provinces), after two years of prosperity in the Wilhelmine Empire, but also after the Berlin stock market crash of 1873. This, historian Harry Liebersohn notes, ended a short period of, 'national high living and careless speculation,' ruining people in all classes, many turning to anti-Semitism and agrarian pressure group politics.[44] Germany was a recently unified but culturally and religiously divided society in economic crisis.

Tönnies saw class interests as extending from a *natural* self-interest, drawn from his reading of Thomas Hobbes' *Leviathan* (1651). Hobbes wrote during the Commonwealth produced by the English Revolution, and argued that only a strong state could avert the chaos of a war of all against all, as the ground on which social relations are built. A state of nature is a violent death but human industry and the pursuit of a common good require an overarching state to provide stability and continuity. Tönnies adopts Hobbes' conflict-ridden scene, but sees the modern state, not as a necessary creator of equilibrium, but now as supporting the class interests of the owners of capital through selective rationality. Liebersohn summarises,

> Hobbes argued that the transition from chaos to civil order took place because reason persuaded human beings to sacrifice part of their freedom out of self-interest. Yet the peace resulting from this consensus, the peace of civil society, nurtured an increasingly destructive form of rationality. In the capitalist marketplace the exercise of reason became a continuation of civil war by other means.[45]

In attempting to avoid this destructive attitude, Tönnies saw the location of community as elusive, in remote pasts informed by philologist and anthropologist Johann Jakob Bachofen's *Das Mutterecht* (1861). Bachofen supposed an archaic matrilineal society as a foundational, pre-classical state of society. For Tönnies, this meant that, since there had once been such a society, it could re-emerge. Liebersohn explains:

> Tönnies today has the reputation of a Germanophile thinker whose yearning for community led away from enlightened Western Europe and down the isolated path to disaster [in the 1930s]. His actual intentions were more nearly the opposite. Few thinkers of his time read and wrote with a more cosmopolitan disregard for national boundaries. ... His historical writings ... represent not so much a divergence from the Enlightenment as an attempt to rethink it under new conditions, anticipating the Frankfurt School's self-criticism of enlightened reason.[46]

Bachofen fancifully proposed a trajectory of four ages—from wild nomadic to matriarchal, to Dionysian transition from matriarchy to patriarchy, and finally to an Apollonian solar cult which formalises patriarchy for ever. Despite the sketchy evidence, he saw this as a scientific social theory. Tönnies relied on Bachofen's historical knowledge, such as it was, because he wanted his theory to be recognised as scientific, too. As to the reference to the Frankfurt school, I agree: Tönnies' aim to revise rationality is echoed, but coincidentally, in *Dialectic of Enlightenment* (1944), a critique of rationality by Theodor Adorno and Max Horkheimer in the shadow of the rise of fascism.

Adorno and Horkheimer, citing the empiricism of Francis Bacon in the 1600s, characterise scientific rationality as domination: 'The human mind ... is to hold sway over a disenchanted nature. Knowledge, which is power, knows no obstacles: neither in the enslavement of men nor in compliance with the world's rulers.'[47] The world is disenchanted—liberated from the rule of a mysterious Fate—opening a path towards a *critical* rationality which transforms the regime of power-over into a sphere of power-to.

Gemeinschaft und Gesellschaft was published two years before Camillo Sitte's *City Planning According to Artistic Principles* (1889), a key text in modern planning in its German original and English translation,[48] and an influence on Garden City architect Raymond Unwin in the 1900s.[49] Urban historian Helen Meller argues that for Sitte, planning continues the project of civilisation: 'the modern city needed to grow from the old.'[50] This conveys a discontent with modern city life but, Meller adds, for Sitte, planning could express individuality, and might facilitate a changing relationship with citizens, affirming society, not community, as the future. As an architect Sitte, 'saw this relationship in terms of an aesthetic approach to town planning.'[51] Meller quotes Sitte:

City planning represents the fusion of all the technical and creative arts into a great and integral whole; city planning is the monumental expression of civic spirit ... [It] regularises traffic, it provides healthy and comfortable living conditions for modern man [sic] ... it has to arrange the most favourable placement of industry and business, and it should foster the reconciliation of class difference.[52]

Sitte's support for regulation and the reconciliation of class differences in modern cities (not in a fabled past) departs from Tönnies' idea of transaction; but Sitte, like Tönnies, wanted to be scientific, basing his aesthetics on observations of city form. An outcome was the design of suburbs as sites for whole but localised societies, with shops, parks and civic buildings.

Tönnies' dualism of community-society is, like Sitte's formulation of a complete theory of planning, a meta-narrative. The transition from village to city entails liberation from ties to the land and the family, at the price of community; society allows the forging of voluntary bonds of shared interest but continuity is lost. Although Tönnies presented his work as an objective record, his sympathy was with the rural, where he lived.

A METROPOLITAN IMAGINARY

The urban imaginary based on the dualism of community-society, which tends towards old rural-new urban, is extended by Georg Simmel, writing in Berlin in the 1900s. In *Metropoles and Mental Life* (1903) he contrasts the constancy of rural life to the shifting stimulations of modern, metropolitan life, and identifies a metropolitan state of mind, the blasé attitude.[53] That is, fleeting impressions and transient relations erode a fixed personality, producing a purposefully constructed self which reflects its shifting surroundings. As Donald puts it, this, 'manifests itself in an aesthetic self-creation.'[54] This is not social atomism, nor the isolation proposed by Tönnies, but a flux in which individuals float voluntarily if at times uneasily in metropolitan air. Donald adds that Simmel does not see, 'a power that oppresses,' but, 'how individual agency is enacted within the field of possibilities realised by this real-imagined environment' in the metropolis, which is both, 'inescapably strange' and yet makes, 'the fabric of our liberty.'[55] Further, in 'The Psychology of Money' (1899), Simmel argues that the money economy pervades society and culture, reducing life to transactions. Being abstract and fluid, it is a suitable metaphor for the mentality of modern urban life. This is expanded in *The Philosophy of Money* (1900), an analysis of the money economy with an emphasis on consumption. The consumer fetishises the goods on display, lending things an emotive force beyond monetary value or use. The outcome is a society of veils, of ends reached by indirect paths whereby the means of negotiation—money, and its accumulation—dominate. Ambivalence emerges in a wavering between desire and possession, or having but destroying the object of desire.[56] In the dichotomy of want and denial, the relation of means to ends is reversed: the means, or the purchasing power which is endlessly convertible, becomes itself the object of desire: so while commodities seem abstract and substanceless money is, 'the concrete and immediate source of pleasure.'[57]

COLLIDING DREAMS

The context for Simmel's thinking is instructive: Berlin in the 1900s. Crowds surged in streets lined with shop windows in which they watched their reflections as well as seeing the goods on display, tramcars sped by, streetlights lit the night, newspapers were published several times a day and cinemas delivered fantasy as well as newsreels. Simmel saw large cities as supporting an advanced division of labour and diversity of services; individuals were pushed to specialise by being in competition with so many others so that, 'city life has transformed the struggle with nature for livelihood into an inter-human struggle for gain.'[58] All this made possible a new cosmopolitanism in which everyone is a stranger in a city of migrants.

After the eighteenth-century bourgeois city and then Romanticism, Simmel says,

> Another ideal arose: individuals liberated from historical bonds now wished to distinguish themselves from one another. ... It is the function of the metropolis to provide the arena for this struggle and its reconciliation ... The metropolis reveals itself as one of those great historical formations in which opposing streams which enclose life unfold, as well as join one another with equal right.[59]

But the *modern* city is unstable, continuously changing and uncertain, lending the village of popular as well as sociological imagination a nostalgic continuity which compensates for the city's uncertainties. Or so it seems. Because Simmel continued to live in Berlin's fashionable West End, enjoying the metropolis. To make a perhaps too sweeping generalisation, Tönnies says that city life is impoverished by the loss of community; Simmel says that metropolitan life, as an advanced state of what Tönnies called society, is liberating, but that its stresses generate a blasé attitude. In a metropolitan city's kaleidoscopic realities, individuals feel new feelings and experience new sensations, swimming forever in a pool's deep end.

Wilson cites Simmel and the Chicago urbanists as emphasising the 'impersonality' of city life[60]; but I think, speculatively, there is a link between Simmel's sense of simultaneity in the metropolis and what Wilson says near the end of *The Sphinx in the City*, citing Sennett's *The Uses of Disorder*:

> Richard Sennett was right to grasp the nettle of disorder, and to recognise that the excitement of city life cannot be preserved if all conflict is eliminated. He was right to emphasise the positive aspects of conflict, and to understand that life in the great city offers the potential for greater freedom and diversity than life in small communities. This is particularly important for women.[61]

She continues that planning remains necessary—there is no regression to Hobbes' state of nature—but that it needs to evolve towards a new purpose:

> Hitherto, town planning has too often been driven by the motor of capital-ist profit and fuelled by the desire to police whole communities. Planners considered it desirable to provide a civilised standard of housing for the masses, but this was less as a right than because they imagined that better housing would lead to a docile and domesticated populace. They wished to eliminate not just dirt and disease, but slovenly housewives and rioting workers. ... The purpose of the plan was to create a city of order and surveil-lance rather than one of pleasure and opportunity.[62]

Planning, like rationality, needs to be revised from within. I quote Wilson at length above (and again in Chap. 6) because she states the mat-ter succinctly, and better than I could paraphrase. But I want to add that the ambivalence of excitement and uncertainty which permeates attitudes to modern cities creates its own problems in fostering Utopian visions of socio-spatial ordering, and dystopian prospects of decline. Both are unreal.

The civic order engineered by design is Utopian only as a design, its status not far removed from fiction; the chaos of inner-city streets seen by the professional observer is not felt by occupants for whom *their* streets house multiple uses. In the 1960s, the dystopian narrative was affirmed by the terms planning blight and concrete jungle but the gap between Tönnies' and Simmel's theories opens a possibility for a more liberating understanding of city life.

On the one hand is the appeal to village community, in sociology and in literature, in suburbia and in New Urbanism, a nostalgic view based on conflict-free illusions. It is Utopia as non-place, not as happy place (*eu-topia*) because it cannot exist given the claims and counter-claims of city living. On the other hand is a recognition of engagement, and of excite-ment. This requires working towards an equilibrium which is contingent, a city which is never completed. On this, Sandercock argues for an agonis-tic politics whereby claims for space and visibility are constantly renegoti-ated, and for a, 'broad social participation in the never completed process of making meanings and creating values.'[63]

As said at the outset, the meanings of words are produced through regulation and usages. The meaning of, not just the word but the experi-ence, of city is mutable. The production of meaning is thus integral to the production of spaces and power-relations. If Tönnies seeks a return to village community, Simmel looks to metropolitan plurality; conditions overwhelm the citizen in a simultaneity of sights and meanings; but this is a supercharged site of human inter-perceptions.

I end by saying that cities bring large numbers of people into close habitation (proximity); they mix with people from different backgrounds (diversity); they enter voluntary ties of common interests, and move through social levels as well as geographical spaces (mobility); and, through solidarity and by enacting shared values, they may gain agency.

If this is Utopian, it is *eu*-topian despite engendering a collision with the dream of a conflict-free condition, and implies an urban imaginary looking to cities as sites of potentially free and fulfilling vitality. At the same time, the conditions in which voluntary ties are entered into are uneven, often favouring those in power or who hold wealth (increasingly now, the same). There never was a really-existing free mixing of strangers in common spaces—in early societies, strangers were regarded with suspicion, often killed—nor a public sphere of shared, free co-determination. This does not invalidate the idea of a public sphere as an extension of the horizon of the possible, but requires a realist, unsentimental critique of cities (plural) as they are and as they become, and discarding of universalising images of either dystopian dis-ease or nostalgic idyll.

In the next chapter I investigate the development of a rural myth from early modern to late modern times. In Chap. 3 I look to the philosophical basis of modernity in René Descartes' philosophy of the subject, and the gesture of drawing lines as regular places. In Chap. 4 I move to what I see as the contradictions of modern architecture and planning. This critique of modernist wastelands leads me to the ruinscapes of post-industrial sites in Chap. 5; and a return to some of the ideas rehearsed above in Chap. 6, extended into brief examination of cases which show the possibility of what French Marxist sociologist Henri Lefebvre calls the Right to the city. Lefebvre is highly critical of Cartesianism (as I discuss in Chap. 3); and I base part of my critical analysis on his theory. I think, too, of my own subjective experience in walking around cities, mainly in Europe, concluding that the presiding image of the city in redevelopment schemes is destructive and dehumanising. There must be other ways to deal with urban problems, because alternatives are always possible; this requires an alternative imaginary, and evidence from actuality that elements of the imaginary can be realised.

NOTES

1. *Shorter Oxford English Dictionary* (1973) Oxford, Oxford University Press, p. 2439.
2. Kohn, M. (2003) *Radical Space: Building the House of the People*, Ithaca (NY), Cornell University Press, p. 22.
3. Benhabib, S. (1986) *Critique, Norm, Utopia: A Study of the Foundations of Critical Theory*, New York, Columbia University Press, p. 34.
4. Wark, McK. (2011) *The Beach beneath the Street: The Everyday Life and Glorious Times of the Situationist International*, London, Verso, p. 38.
5. Mould, O. (2017) *Urban Subversion and the Creative City*, Abingdon, Ashgate, p. 113.
6. Sennett, R. (1995) *Flesh and Stone: The Body and the City in Western Civilisation*, London, Faber and Faber, pp. 195–196.
7. Soja, E. (2000) *Postmetropolis: Critical Studies of Cities and Regions*, Oxford, Blackwell, p. 324.
8. Soja, *Postmetropolis*, pp. 36–42; Mellaart, J. (1967) *Çatal Huyuk: A Neolithic Town in Anatolia*, London, Thames and Hudson.
9. Mellaart, *Çatal Huyuk*, p. 67.
10. Gold, J. (2008) 'Modernity and Utopia, 'Hall, T., Hubbard, P. and Short, J.R., eds., *The Sage Companion to the City*, London, Sage, p. 80.
11. Sennett, *Flesh and Stone*, p. 155.
12. Soja, *Postmetropolis*, p. 248.
13. Sennett, *Flesh and Stone*, p. 157.
14. Donald, J. (1999) *Imagining the Modern City*, London, Athlone, p. 27.
15. Engels, F. (1892) *The Condition of the working Class in England in 1844*, London, Allen and Unwin.
16. Miles, M. (2019) *Cities and Literature*, London, Routledge, pp. 45–65.
17. Lehan, R., (1998) *The City in Literature: An Intellectual and Cultural History*, Berkeley, University of California Press, p. 211.
18. Spengler, O. [1918–22] (1926) *The Decline of the West*, New York, Knopff, p. 32, quoted in Lehan, *The City in Literature*, p. 211.
19. http://35percent.org/heygate-regenrtration-faq - see also www.heygatewashome.org [both accessed 8 January 2020].
20. Walker, P. (2013) 'Bailiffs will sound death knell for vast London estate,' *The Guardian*, 5 November, p. 18.
21. Tinker, T. (2011) quoted in Moss, S., 'Homes under the hammer,' *The Guardian*, 4 March, section 2, p. 9.
22. Sennett, *Flesh and Stone*, pp. 334–338.
23. Wilson, E., 1991, *The Sphinx in the City: Urban Life, the Control of Disorder, and Women*, Berkeley (CA), University of California Press, p. 46.

24. Robins, E., 1996, 'Thinking Space / Seeing Space: Thamesmead Revisited,' *International Journal of Urban Design*, 1, 3, pp. 283–291.
25. MacCannell, D. (1999) 'New Urbanism and its Discontents,' Copjec, J. and Sorkin, M., eds., *Giving Ground: The Politics of Propinquity*, London, Verso, pp. 106–130.
26. Wilson, *The Sphinx in the City*, p. 107, citing Rothblat, D., Garr, D. and Sprague, J. (1979) *The Suburban Environment and Women*, New York, Prager, p. 13.
27. Geddes., P. (1915) *Cities in Evolution: An Introduction to the Town Planning Movement and the Study of Civics*, London, Williams and Northgate, p.219, quoted Wilson, *The Sphinx in the City*, p. 106.
28. Wilson, *The Sphinx in the City*, p. 112.
29. Sennett, R., 1996, *The Uses of Disorder: Personal Identity and City Life*, London, Faber and Faber, p. 8.
30. Sandercock, L. (1998) *Towards Cosmopolis*, Chichester, Wiley, p. 192, citing Sennett, *The Uses of Disorder*, p. 92.
31. Sennett, *The Uses of Disorder*, p. 72.
32. Sandercock, *Towards Cosmopolis*, p. 194.
33. Heward, E. (2019) 'The urban village beneath the Mancunian Way,' *Manchester evening News*, 16 June [accessed on-line https://manchester-eveningnews.co.uk/whats-on/whats-on-news/hatch-mancunian-way-manchester-flyover-16433989 accessed 22 January 2020].
34. Tottle, J., quoted in Heward, 'The urban village beneath the Mancunian Way' [accessed on-line as above].
35. Gratz, R.B., 2003, 'Authentic Urbanism and the Jane Jacobs Legacy,' Neal, P. ed., *Urban Villages and the Making of Communities*, London, Spon, p. 28.
36. Jacobs, J.,(1961) *The Death and Life of Great American Cities*, New York, Vintage, p. 168.
37. Sennett, *Flesh and Stone*, p. 357.
38. Sennett, *Flesh and Stone*, p. 358.
39. Wright, G. (2002) 'Permeable Boundaries: Domesticity in Post-war New York,' Madsen, P. and Plunz, R., eds., *The Urban Lifeworld*, London, Routledge, p. 209.
40. Neal, P. (2003) 'An urban village primer,' Neal, *Urban Villages*, p. 2.
41. Neal, 'An urban village primer,' p. 27.
42. Tönnies, F., [1887] (1957) *Gemeinschaft und Gesellschaft*, [English translation] East Lancing (MI), Michigan State University Press.
43. . Frisby, D. (1985) Fragments of Modernity, Cambridge, Polity, p. 46 citing Tönnies, F. (1895) 'Considérations sur l'histoire modern,' *Annales de l'institut international de sociologie*, 1, pp. 245–252 [esp. p. 246].

44. Liebersohn, H. (1988) *Fate and Utopia in German Sociology, 1870–1923*, Cambridge (MA), MIT, p. 14.
45. Liebersohn, *Fate and Utopia*, pp. 22–23.
46. Liebersohn, *Fate and Utopia*, pp. 26–27.
47. Adorno, T.W. and Horkheimer, M., [1944] (1997) *Dialectic of Enlightenment*, London, Verso, p. 4.
48. Sitte, C., [1889] (1986) *City Planning According to Artistic Principles*, trans. Collins G R and Collins G W., London, Phaidon.
49. Miller, M. (1981) 'Raymond Unwin,' Cherry, G.E., ed., *Pioneers in British Planning*, London, Architectural Press, p. 84.
50. Meller, H. (2001) *European Cities 1890–1930s: History, Culture and the Built Environment*, Chichester, Wiley, p. 42.
51. Meller, *European Cities*, pp. 42–43.
52. Sitte, *City Planning*, p. 190.
53. Simmel, G., [1903] (1997) 'The Metropolis and Mental Life,' Frisby, D. and Featherstone, M., eds., *Simmel on Culture*, London, Sage, pp. 174–185.
54. Donald, *Imagining the Modern City*, p. 11.
55. Donald, *Imagining the Modern City*, pp. 11–13.
56. Liebersohn, *Fate and Utopia*, p. 135.
57. Liebersohn, *Fate and Utopia*, p. 135.
58. Simmel, 'The Metropolis and Mental Life,' p. 180.
59. Simmel, 'The Metropolis and Mental Life,' p. 185.
60. Wilson, *The Sphinx in the City*, p. 130.
61. Wilson, *The Sphinx in the City*, p. 156.
62. Wilson, *The Sphinx in the City*, p. 156.
63. Sandercock, L. (2006) 'A Love Song to our Mongrel Cities,' Binnie, J., Holloway, J., Millington, S. and Young, C., eds., *Cosmopolitan Urbanism*, London, Routledge, p. 50.

CHAPTER 2

From Arcadia to Plotlands

Abstract This chapter traces the evolution of an Arcadian myth from the sixteenth to twentieth centuries. It begins in Port Sunlight, a model industrial village in Merseyside built in the 1900s, which evokes a dream of rural community enabled by enlightened paternalism. But that dream began as an aristocratic retreat from court life in Elizabethan England, at the Earl of Pembroke's estate at Wilton where Philip Sydney wrote his poem Arcadia (1588). The chapter moves to the eighteenth-century English landscaped park, taking Rousham, Oxfordshire as an example; and the Garden City movement of the 1900s, then a grass-roots pastoralism in working-class rambling, and in rural and coastal settlements (plotlands).

Keywords Arcadia • Rural idyll • Landscaped park • Suburbia • Garden City • Plotlands

In Chap. 1 I moved from discussion of the terms city and urban to attitudes to the rural (and community) in early sociology. In this chapter I look at the myth of a rural idyll, used to counteract the effects of city living. This begins, in English history, at Wilton near Salisbury, where the Earls of Pembroke created an Arcadia on their estates. It reappears in a more formalised way in the English landscape, and it informs model industrial villages such as Port Sunlight in Merseyside. I begin the chapter at Port sunlight, and an intersection of a rural myth and modern industrial

© The Author(s) 2021   23
M. Miles, *Paradoxical Urbanism*,
https://doi.org/10.1007/978-981-15-6341-6_2

life. I trace the myth through aspects of twentieth-century urbanism, and note its adaptation to the plotlands of working-class escape from the city.

PORT SUNLIGHT

Port Sunlight, near Birkenhead, across the river Mersey from Liverpool, is a model industrial village for workers at the Sunlight Soap factory, commissioned by William Lever (later Lord Leverhulme). Around 800 houses were built from 1899 to 1914 on a marshy site acquired to expand the Lever Brothers soap business under the brand Sunlight Soap. Port Sunlight was a model of progressive industrial housing but it was based on an idea of the English village, the result of progressive but patriarchal entrepreneurship.

For Sunlight Soap workers it offered indoor bathrooms, hot water on tap, and unshared toilets, major advances on standard working-class housing. Each house had a garden, and the streets had green verges in a density around a tenth of the typical 100 to 130 houses per acre of industrial towns at the time. There were social clubs, a theatre, and allotments for growing vegetables and fruit. Port Sunlight was described as 'exceptional' in *Building News* in 1899.[1] Lever Brothers were progressive employers, paying higher wages for shorter hours than their competitors, with an incremental pay structure which enabled women to build careers, well before they were given the vote. But the housing provided at fair rents was tied to continuing employment, implying a requirement for good behavior, and women had to leave work on marriage.

Addressing the International Housing Conference, which visited Port Sunlight in 1907, Lever said,

> The building of ten to twelve houses to the acre is the maximum that ought to be allowed ... Houses should be a minimum of 15 feet from the roadway ... every house should have space available in the rear for a vegetable garden. Open spaces for recreation should be laid out at frequent and convenient centres ... A home requires a greensward and garden in front of it, just as much as a cup requires a saucer.[2]

The appreciation of art, literature and music was encouraged at Port Sunlight, and the Lady Lever Art Gallery still offers free admission to the paintings, ceramics and furniture collected by William and Lady Lever. Initially only a temperance (alcohol-free) hotel was provided but villagers requested beer; after a vote in 1903 (open to women as well as men) the Bridge Inn served alcohol. The Victorian gothic Christ Church (1902–04) is non-denominational. Edward Hubbard and Michael Shippobottom comment on this as, 'difficult to imagine outside the special circumstances and rarified atmosphere of Port Sunlight,' while its style expresses Lever's, 'love of medieval churches … [and] desire for beauty and dignity in worship and liturgy.'[3]

English Arts and Crafts architecture of the 1890s informed Lever's choice of architects. Each housing block is unique in design within an overall gabled cottage style. Most houses have half-timbered façades in red brick, with decorative chimney stacks. Architects included the firms William Owen, Douglas and Fordham, Grayson and Ould, and J. J. Talbot, from North-West England; but also the more prominent Edwin Lutyens.

The terraces which line Park Road, two blocks by Owen, another by Fordham, opposite the park, the Dell, are typical of Port Sunlight's allusion to Tudor England as an image of stability and continuity (Fig. 2.1).

Fig. 2.1 Port Sunlight, Merseyside. (Author's photograph)

This is Shakespeare's England as antidote to industrialisation. In fact, the Elizabethan England they conjure was marked by Papist plots and fear of invasion, and by malnutrition in rural areas after the destruction of the monasteries which had sustained the poor and vagrant. Of course, in the 1890s, when Park Road was built, the allusion to Elizabethan times was illusory, conveniently ignoring the violence associated with Tudor England. Against industrial blight, an Arcadian world offers the tranquility found, too, in Alfred Tennyson's medievalist narrative poem *Morte d'Arthur* (1857),

> But now farewell. I am going a long way…
> To the island-valley of Avilion;
> Where falls not hail, or rain, or snow,
> Nor ever wind blows loudly; but it lies
> Deep-meadow'd, happy, fair with orchard lawns
> And bowery hollows crown'd with summer sea,
> Where I will heal me of my grievous wound.[4]

Port Sunlight was not Avalon (identified with Glastonbury in Somerset), but a philanthropic venture aimed at improving the quality and length of life of Sunlight Soap workers and their families. The wound it sought to heal was that inflicted by industrialisation. Lever supported the reforming Prime Minister William Gladstone, and was a Liberal Member of Parliament from 1905 to 1909. He was called a radical, and Port Sunlight is among the most progressive experiments of nineteenth-century housing design, alongside Bournville near Birmingham.

Like Port Sunlight, Bournville, built by the Cadbury brothers for workers at their chocolate factory, had houses in small, distinct blocks of varied design. When built in 1895 there were no indoor bathrooms, but these were later added. According to a guide book, 'Bournville showed how cheerfulness could be brought back into the industrial suburb … Sunlight and fresh air could find their way freely into the houses, and even the kitchens and sculleries were bright and attractive.'[5]

I do not want to detract from the radical stance involved in providing humane conditions for factory workers, but the Lever and Cadbury enterprises sought to engineer a social order via paternalism. Although Lever saw the personal life of employees as irrelevant, 'provided the man [sic] is a good workman,' he cautioned that the wrong kind of behavior might threaten a tenancy.[6] Good behavior meant sobriety, which ensured higher productivity, but also education. This was promoted by an illustrated almanac, the Sunlight Soap Year Book, with practical advice and entries on geography and the Empire, science, literature, art and sport, and cycling (with free maps). Copies were distributed to head teachers (to the annoyance of competitors) while District Agents spread the word overseas. The houses were sold in the 1980s following legislative changes regarding tied houses; and few workers now live in Port Sunlight, the village's listed status preventing house extensions and adaptations.

ARCADIA

Port Sunlight is an industrial era reversion to a pre-industrial form of settlement, with new conveniences such as hot water. The historical image which informs it can be traced to Tudor England, when the first Arcadia was created at Wilton, near Salisbury. This followed both classical precedents such as Virgil's Latin pastoral poems, and their renewal in early modern forms in the sixteenth century.

Virgil's poetry was known among the literate classes in sixteenth- and seventeenth-century Europe. Among Renaissance literary origins of the Arcadian myth are Jacopo Sannazaro's poem *Arcadia* (1504) and Philip Sidney's *Arcadia* (published as *The Old Arcadia*, 1580) which he wrote while a guest at Wilton. Literary historian M. J. H. Liversidge points out that Virgil's pastoral world gains a new understanding through the illustrations to Renaissance editions of his work, and that these are, 'one of the principal starting points for landscape painting as an independent genre in European art.'[7] And from art and poetry the image is transposed to the design of landscapes (itself a modern idea).

The landscape at Wilton was extensively reconfigured by the first Earl of Pembroke, with the demolition of buildings and enclosure of what had been common land. For the Earl, this was a retreat from the frenetic ambience of court life in London, separated from it by distance and by a mythic temporally, in a pastoral realm. The geographical Arcadia was mythicised in classical Greece as a site of rural bliss ruled over by the flute-playing Pan. Now it was made anew in a landscape of woods and chalk downs. But it deprived the villagers of grazing rights and foraging for wood, and destroyed a pattern of land use established in Saxon times, all as an unquestioned aristocratic privilege (in the eyes of the aristocracy).

Writer Adam Nicolson reads the desire to retreat from politics as an abiding trait among the English aristocracy, dreaming of, 'a lost world, an ideal and unapproachable realm of bliss and beauty.'[8] Introducing Sidney's poem, Katherine Duncan-Jones observes, 'At moments Arcadia is presented as a place far away and long ago ... a world where archetypes of all that is good and bad in contemporary England are to be found in primitive form; but more often ... the story is vividly immediate.'[9] Sidney's *Arcadia* is arranged in five books with sections of prose and verse, partly in the form of dialogues between characters in a comic romance of courtly love. Demonstrating a blurred edge between late medieval and Renaissance culture, the poem echoes the literary form of Chaucer's *Troilus and Criseyde* (c.1385), hinging the plot around the middle section. *Troilus and Criseyde* is set in the Trojan wars but Sidney sets his poem in an archaic Greece, which is not Homeric Greece but a notional realm mediated by Renaissance literature. Sidney read parts of *Arcadia* to his sister, Lady Pembroke.[10] He describes Arcadia as known, 'for the sweetness of the air and other natural benefits. But principally for the moderate and well-tempered minds of the people.'[11] The Arcadians are content and quiet. The Muses bestow perfections. Shepherds, 'had their fancies opened to so high conceits as the most learned of other nations have been long time since content both to borrow their names and imitate their cunning.'[12] This is the backdrop to a romance: 'that wonderful passion which to be defined is impossible, by reason no words reach near to the strange nature of it. They only know it which inwardly feel it. It is called love.'[13] The plot is complicated, and outside my scope here; enough to say that Sidney's *Arcadia* renews a genre of courtly love in context of an aristocratic retreat into private life, just as Virgil wrote agrarian poetry in a time of Roman civil wars.

Arcadia never gained a final form, a revised version remaining unfinished at Sidney's death in 1587. It is an aristocratic dream. Nicholson remarks,

> The world of the Pembrokes was one which none of us could tolerate now: it put the claims of social order far above any individual rights; it considered privacy, except for the highly privileged, a form of subversion; it was profoundly hierarchical and did not consider either people or the sexes equal.'[14]

But it also refused market forces, when trade had already begun to define the aims of a merchant class. In this vision, the poor were worthy of pity on condition of retaining their allotted role. Arcadia, then, is a

Fig. 2.2 Wilton, the park seen from the lawn. (Author's photograph)

pre-capitalist world in which the aristocracies enjoy a freedom to inscribe their dreams on their land, regardless of any consequences for others.

Despite its natural appearance, the park at Wilton (Fig. 2.2) is no more natural than the house which replaced the medieval abbey on the site. Retention of a dovecote, forge, grange (for grain payments in kind) and mill imply a continuity but the village of Washern was demolished. Villagers were given reductions in feudal dues but their just-viable agricultural system was broken. Part of the estate was a royal hunting ground in Saxon times, where red deer roamed; but, as Nicholson continues, 'all of this was a life out of time, shepherds and shepherdesses appearing now just as they had in Virgil's Eclogues and the Greek portrayals of Theocritus, the foundation levels of the Arcadian dream.'[15] Wilton became a reiteration of the classical Greek separation of practical knowledge (utility) from the higher (leisured) realms of philosophy and poetry, while, Nicholson continues,

Most of rural England, from the Middle Ages onwards, spent most of the time under stress. There was a desperate shortage of fertility; farming systems could only just sustain the human populations that depended on them. If for every grain sown, the average return was between three and four grains, one of which had to be kept as seed corn for the next year, the land was an asset to be contested. Nothing could be allowed to disrupt the habits which ... had allowed the village to feed its people.[16]

In 1549 the people of Washern tore down the fences round the new enclosures and killed the deer. William Herbert was in Wales but returned with an armed retinue. They killed the villagers as if hunting. Nicholson writes,

If you stand outside Wilton House today staring across the elegance of the park ... you are looking at one of the heartlands of Arcadia: a stretch of landscape in which the people who claimed some rights over it were murdered so that an aesthetic vision of an otherworldly calm could be imposed in their place.[17]

This pastoral vision, soaked in blood, was the first English landscaped park. The green vistas on which a village stood are now a tourist site, masking the origins of the English pastoral in a violent assertion of power over others and nature.

THE LANDSCAPED PARK

The eighteenth-century English landscaped park may seem to be natural but is as managed as any formal garden. For garden designers such as William Kent and Capability Brown, the natural growth of trees and grass was a resource to be used like the colours on an artist's palette, influenced by the paintings of Claude Lorrain which were collected by the landed gentry. These pictures used a balanced asymmetry; and landscape designers worked in an equivalent but three-dimensional way, juxtaposing wooded vistas to closer sights of bridges, fountains, grottoes and classical-style temples, on a route from the house through open and wooded areas, up and down low hills, back to the house by another route. In the park, all is tranquil and orderly, and contained. Yet nature's cultivation is also an outcome of modern social and scientific modes of ordering, as architectural historian Jonathan Hill writes,

As nature was considered a machine, mankind could have been its driver and engineer, making technical adjustments to improve performance. But in an era that associated power and status with land ownership ... the gentleman farmer was a model for the enlightened management of nature and society.[18]

Gentlemen farmers—the landed gentry—made improvements to land use and drainage, bought higher-yield seeds, and extended the enclosure of common land.

In the eighteenth century, the similarity of a park to the painted landscapes on the walls of a grand house denoted aristocratic ease and an educated mind. Being beautiful, a park is as useless as a work of art, an object of contemplation. But, as non-productive land, it is a sign of wealth, and this wealth was the result of new controls on grain imports and prices, from 1773 through to the Corn Laws in 1815 (repealed in 1846). Grain profits contributed to the cost of parkland, along with revenue from mineral extraction and colonial ventures such as sugar (a slave economy). Against this background of ruthless exploitation, the park evoked a temporality between a classical past and a present set aside from the tribulations of politics, social unrest and foreign wars. As at Wilton earlier, whole villages were cleared. Thomas Coke, owner of the park at Holkham in Norfolk, said, 'It is a melancholy thing to stand alone in one's own country ... not a house to be seen but for my own. I am Giant of Giant's Castle, and have ate up all my neighbors.'[19] Melancholy is a low price for others' abjection.

Kent made several drawings for Holkham in the 1730s. In one, a gondola carries a fisherman on a lake made by damning a river, while two figures walk the lawn; in another, two figures ride in a gondola.[20] In this Romantic sublimity, intimations of mortality are distanced; and, as Hill writes, 'the picturesque garden was equivalent to a history, formulating an interpretation of the past in the present ... [and] as ancient Rome was a model for Georgian Britain, classical forms were ... simultaneously ancient and modern.'[21]

ROUSHAM

The park at Rousham, designed by Kent between 1738 and 1741 for General James Dormer, looks like a natural landscape, contrasting to the formality of the walled garden around the seventeenth-century house. But this is the picturesque, derived from a mix of Renaissance, Oriental and

Italianate sources, presenting an image of a society prosperous enough to set aside non-productive land. The park's privileged strollers are like actors on the proscenium stage, able to make decisions which shape the plot as they follow the script of vista and incident, classical allusion and natural distance. Both entail a suspension of disbelief, in the theatre for the audience, in the park for the wealthy owners and their guests as they look on a drama of temples, monuments and fake ruins in a seemingly timeless realm. So, the Vale of Venus—two cascades with rustic (Palladian) stonework, a statue of Venus flanked by two swans—is both Roman and Georgian English, one Empire echoing another. To the side is a vaulted arcade, the Praeneste, flanked by classical vases (Fig. 2.3).

Fig. 2.3   Rousham, the Praeneste. (Author's photograph)

Rousham is polyvalent in references, just as a privileged education included multiple aspects of the past. To the west is a classical temple, to the east a pyramid-roofed house and to the north an Italian gothic house. The pyramid does not resemble Egyptian pyramids, predating excavations there; Kent's source is an illustrated Renaissance romance—Francesco Colonna's *Hypnerotomachia Poliphili* (1499, *The Strife of Love in a Dreame*, 1592).[22] But Kent uses his sources eclectically: the park reconstructs history in sequences of visual incident among a mix of native and imported trees. But this is aesthetic space, separated from actuality. Yet, as geographer Stephen Daniels argues, there was a moral purpose as well,

> The idea of improvement was central to landed culture. Initially used to denote profitable operations in connection with land, notably aristocratic enclosure, by the end of the [eighteenth] century improvement referred to a range of activities. It denoted moral and aesthetic dimensions or implications of estate design and management; furthermore it referred to a broad range of conduct, from reading to statecraft.[23]

Daniels adds that while the landed gentry saw coherence in a landscaped vista, this was a coded image of a holistic society, while, 'improvement implied a progressive, stable polity ... in an increasingly complicated society.'[24] Outside were bread riots and military campaigns.

By the nineteenth century improvements included cutting down trees and driving roads through meandering valleys. For Daniels, 'The new regime is as uninviting for the spectator as it is for the labourer,' while land is, 'so ruthlessly mobilised for money' that it is no longer landscape.[25] As enclosures increased and workers were evicted from their houses due to mechanisation, unrest grew, exacerbated by artificially high grain prices. Unrest was often supported by ordinary people. For example, at Otmoor near Oxford in 1829, villagers took down seven miles of fences; forty-four people were arrested but when they were taken to Oxford, a crowd gathered in the streets for St Giles Fair freed them.

MORTALITY AND LIBERALISM

For cultural historian Simon Pugh, Kent's design at Rousham illustrates the pastoral vision derived from classical literature which, while not like the Roman *campagna* it evokes, did possess, 'those universal ingredients which give it identity with that landscape.'[26] This draws on the classical cultural framework provided in elite education and the Grand Tour on which the gentry saw Mediterranean culture, and on Virgil's *Georgics* and *Eclogues*, including John Dryden's translation. Historian and curator Christopher Woodward notes a fascination with mortality whereby grottos and black rocks reminded visitors of their vulnerability, while the park's naturalness was reassuring. Woodward writes,

> Ladies were expected to shiver with horror as the path disappeared into a cold, dark grotto with a waterfall thundering in the invisible distance. Emerging into a gentle valley grazed by sheep they paused on the steps of a classical temple, and a gentleman in the party might be moved to declaim Virgil's *Georgics*. These country gardens were designed as circular walks deliberately punctuated by such incidents, and in the eighteenth century were opened to all respectable members of the public.[27]

Darkness encountered within the light is licensed shock, like the Sublime which appears in a painted storm. At Rousham, eclectic historical appropriations construct a perpetual present, but also tension between past and present. Pugh writes:

> The simulated regression to a primitive age does nothing to question the permanent crises that were beginning to be recognised as a feature of capitalism. This regression disguises barbaric forms of social domination dressed up in the seeming egalitarianism of a free world of natural justice and rights and a natural habitat.[28]

Mortality is further denoted by a sculpture of a dying gladiator by Peter Scheemaeckers (now a copy). The gladiator states proto-Christian sacrifice, perhaps referring to General Dormer's death in 1741, when the park was completed. But Pugh remarks, aware of the military history of the time, 'the English garden takes over the military images of control, and cleverly redefines those images in an emerging language of liberal democracy.'[29]

Liberalism did not mean mass enfranchisement but maintenance of a stable regime. When an old village was cleared its occupants were rehoused, sometimes in houses of improved quality at the entrance to the estate, setting a precedent for model industrial villages such as Port Sunlight and Bournville a century later. Writer Gillian Darley observes that these houses were in rows: 'instead of the signs of a community which has gradually evolved in an organic fashion ... the cottages are immediately obvious, built from the same materials and placed symmetrically.'[30]

Putting houses outside the gates kept them out of sight from the house; and the occupants were not allowed into the estate without specific purpose. The rows were also evidence of the land-owners interest in improvement. Darley notes that, in the eighteenth century, to be philanthropic was, 'not merely unfashionable, it was eccentric.'[31] The new houses for the poor had classical proportions, however, as instruments of social as well as spatial ordering. At Nuneham Courtenay between London and Oxford, Lord Harcourt cleared a village from a hilltop above his Palladian-style house in the 1760s, replacing its church with a neo-classical temple a mile away and building lines of cottages on the turnpike road. A visitor, Stebbing Shaw, remarked that the houses offered, 'the comforts of industry under a wholesome roof ... by the liberal assistance of his lordship.'[32] But these were vernacular in style, a precedent for Bournville and Port Sunlight, and the Garden City in the 1900s.

THE GARDEN CITY

In the 1900s, the Garden City movement sought to re-construct urbanism in rural terms via pseudo-vernacular design styles in small cities on green-field sites.[33] Urban opportunities for employment were to be merged with rural fresh air, assisted by a depression in rural land values. In Ebenezer Howard's *Garden City* (1898), the new settlement's population is limited to 32,000, after which satellite towns will be built; the built area is 1000 acres, set in 5000 green acres. An adjacent industrial zone includes manufacturing, engineering and jam-making. Garden City is circular in plan, dissected by six boulevards. The centre consists of a town hall, art gallery, museum and concert hall around a park. Howard writes,

> Running around the Central Park is a wide glass Arcade or Crystal Palace. This building is in wet weather one of the favourite resorts of the people; for the knowledge that its bright shelter is close at hand will tempt people into

the park even in the most doubtful of weathers. Here manufactured goods are exposed for sale, and here most of the shopping which requires the job of deliberation and selection is done.[34]

This merits unpacking. Leaving aside an English obsession with the weather, the glass arcade echoes the Crystal Palace of the Great Exhibition, London in 1851, but also the iron and glass galleries which Charles Fourier envisaged housing his Utopian community, the Phalanstery. Fourier, in turn, derived this from Parisian arcades built from the 1820s onwards to enclose luxury shops.

Howard's ideas were drawn from literary sources, too, including Edward Bellamy's Utopian *Looking Backward* (1887), in which Boston in 2000 is a smoke-free city with a productivist, centralised economy; and Peter Kropotkin's *Fields, Factories and Workshops* (1888–1890), an Anarchist tract on self-governing industrial villages. Among built precedents were Port Sunlight and Bournville, built on new sites; but Howard's experience in Chicago, arriving a year after fire destroyed much of the city in 1871, enhanced his attraction to a blank sheet for planning. Howard's idea departs, however, from nineteenth-century Utopian villages such as the Chartist Land Company's 1848 settlement at Minster Lovell in Oxfordshire or the Tolstoyan settlement of Whiteway in the Cotswold Hills, founded in 1898. These were village-scale. Garden City was, as it says, a city.

Garden City was conceptualised via what Howard called three magnets: Town, Country and Town-Country.[35] While Town had negative attributes including the exclusion of nature, high rents, bad housing and foul air, it attracted people through employment and higher wages than those in agriculture. Country attracted people through natural beauty, sunshine, fresh air and clean water, but offered low wages and lack of amusement. Town-Country fused the positive aspects of both, and added the appeal of new houses. For Howard, 'The key to the problem how to restore people to the land' was 'a portal through which ... will be seen to pour a flood of light' on intemperance, toil, anxiety and poverty.[36] Writer Tony Judge notes that for Howard, 'The importance of fresh air, and natural beauty ... was essential if the objective of a better society was to be achieved.'[37] Howard envisaged common ownership of the land, to be held in trust by what he called responsible people until the loan required to pay for it was repaid. He tried to persuade businesspeople to invest, when returns on bonds and securities were low. This was a shrewd marketing device, but it failed when few industrialists were persuaded, and of those who were, Lever left the Board in 1904. No dividend was paid until 1913. Common ownership was abandoned. The outcome was Letchworth and Welwyn Garden City, both becoming middle-class enclaves.

Before moving on, I want to reflect on the glass gallery by the park. This was inspired by the Crystal Palace, re-sited to Sydenham in south London, near Howard's house; and by glassed arcades in London and Leeds, resembling those of Paris. A vogue for winter gardens grew in seaside resorts, a metropolitan equivalent being Alexandra Palace in north London. These were distinctly modern, possibly precursors of more recent shopping malls, or the Winter Garden at Battery Park City in New York. But the Crystal Palace, built for the 1851 Great exhibition, contained 17,000 exhibits from Britain and its colonies, intended to show the British Empire as pioneering the latest industrial processes. Critical theorist Peter Sloterdijk writes of 'a new aesthetics of immersion' in a transition from the world of the middle-class home and department store to a world of, 'magical immanence transfigured by luxury and cosmopolitanism.'[38] In the interior of the Crystal Palace, a world structured by capitalism—productive and innovative—tells the spectator that no other world is required. This is where the end of history begins: conflict is replaced by consumption; social life is 'integrated into a protective shell.'[39] Perhaps this principle of universalisation, a single, modernising model for society, does underpin Howard's proposal. I recall Darley's remark that, 'Life at Port Sunlight still bears the marks of an oppressively paternalistic regime' because everything is provided by the company.[40] Garden City, if built as planned, might have had a similar difficulty in the need for residents to subscribe to its ethos; or perhaps it would have been more liberating, a node of more imaginative housing design and humane working spaces than those which were actually built at Letchworth and Welwyn.

For architects and planners in the 1920s the eighteenth-century represented an aesthetic peak. For instance, the 1928 Countryside and Footpaths Preservation Conference recorded, 'A hundred years ago, before the railway age, this island was almost all beautiful ... not least the improvements that had been made in the eighteenth century, harmonising with the older parts in which they were set.'[41] Geographer David Matless writes, 'the eighteenth century is evoked as an era of improvement and design whose spirit was lost in Victorian times but might be renewed ... Preservationist planning becomes the latest manifestation of an English tradition of progress.'[42]

Looking back is the new looking forward, as if the pre-industrial era was a time of bliss and plenty, of tranquility and universal ease; not of malnutrition, bread riots and the hunting down of dissenting villagers as so much (inedible) meat.

WORKING-CLASS ARCADIAS

Parallel to the development of Town and Country planning in the mid-twentieth century, there is a history of working-class reclamation of the land for leisure. A focal point was the mass trespass at Kinder Scout, Derbyshire, organised by ramblers from Manchester with the support of the British Workers Sports Federation, in 1932. Five ramblers were sent to prison by magistrates who were land-owners and ex-military officers. But the campaign led to the demarcation of National Parks.

A broad growth of working-class outdoor pursuits—walking, hill climbing and cycling—was aided by the Youth Hostel Association's provision of low-cost overnight accommodation in rural places. For Judge, 'the socialist view that the un-spoilt nature of the countryside was somehow morally superior and healthier than urban life' was influential.[43] Perhaps it was an escape from drudgery; perhaps it was also a genuine discovery of the outdoors. Either way, the anti-urban myth of fresh air was a counter to perceptions of grime and moral dereliction in industrialised cities. It informed the Labour League of Youth's summer camps, from 1924 onwards, providing political discussion and solidarity in rural sites, with tents and camp fire singing. In a more extreme form, it contributed to the Kindred of the Kibbo Kift founded by artist John Hargrave in 1920. The Kift held summer camps at which people dressed in fake medieval costumes, danced in fake pagan rites, appropriated tipis and feather headdresses from indigenous North American culture, and borrowed Norse runic writing. Politically it was both progressive—supporting disarmament—and reactionary—advocating free trade—while romanticising Anglo-Saxon history—folk moots—via a green-tinted lens. The Kibbo Kift was as middle- as it was working-class, and had a quaintly English image of wholesome life involving sleeping in tents and going for cross-country walks (in the rain if necessary, to prove something vaguely like resilience).

Hargrave retained control as Head Man but creative activities were delegated to groups named in quasi-Anglo Saxon style as Watlingthing, Wandlething, Aescdalthing, and North Folk, coming from, respectively, North London, South London, South-west London, and the North of

England. The Things organised moots (meetings), wapenshaws (exhibitions) and seasonal hikes. A 1927 illustrated text lists the Kift's points of covenant as:

1. Open air education for the children, camp training and nature craft;
2. Health of body, mind and spirit;
3. Craft training groups and craft guilds;
4. The Woodcraft family, or Roof tree;
5. Local folkmoots and cultural development;
6. Disarmament of nations—Brotherhood of Man [sic];
7. Internal education based on these points, freedom of trade between nations, stabilisation of the purchasing power of money, (in all countries), open negotiations instead of secret treaties and diplomacy, a World Council.[44]

Point 7 seems a catch-all, reflecting several ideas current in England between the wars. In 1928, Hargrave adopted the policy of social credit— a means to alleviate poverty through a rebalancing of the money supply, also advocated by Oswald Mosely, then a Labour politician—in a shift towards the political. Among supporters of social credit in the 1930s were writers T S Eliot, Aldous Huxley and Ezra Pound, and art critic and educationalist Herbert Read.

The Kift became the Green Shirts, and Hargrave asked Pound to write a book of Green Shirt Cantos. He didn't, but sent a song and money for a special flag. The Kift's politics are quirky and ideologically ambivalent; and its antics veer between boy-scout camps and militarism. One of Hargrave's texts, *The Totem Talks* (1923) was translated into German. I mention the Kift because it fits into a wider move to outdoor pursuits and crafts as a search for rural and pre-industrial lifestyles in the inter-war years. For cultural historians Cathy Ross and Oliver Bennett, the attraction for young members was, 'in part the sense of strength, meaning and purpose. If civilization was going downhill fast, joining the Kibbo Kift was joining a crusade for a better future.'[45]

Leaving that aside, working-class rural pursuits re-aligned the rural myth as an alternative to modern, urban lives beset by routine. The construct of the countryside is an urban idea, not a rediscovery of wild nature. And a vehicle for a vision of a better world than that of mills and factories, glimpsed by members of an industrialised society in an economic depression. Judge explains that, while the countryside nourished intellectual and spiritual needs,

> The belief in this effect explains the almost reverential attitude to the country and its indigenous crafts and culture by many early socialists. The exile of the working class from the countryside ... had prevented them fulfilling their potential as whole human beings. They had been cheated of their heritage, and finding the means to re-engage with it was a necessary preparation for a socialist future.[46]

They *were* cheated, in centuries of land enclosures and clearances. But they were deprived of their urban heritage, too, the profits of rent and toil going to the owners of capital.

In the 1920s and 1930s, another kind of reclamation occurred when plotlanders built their own dwellings in rural and coastal sites, sometimes using old railway carriages supplied at low cost by the rail companies (who transported them to the nearest station). Utopian historians Dennis Hardy and Colin Ward observe,

> Irrationally, perhaps, but understandably, the modern plotlanders themselves drew on these historic reasons for settlement. Far removed though they were ... there were still grounds for echoing the original claims. Their little plots ... were still to many a living symbol of freedom and independence.[47]

Part of the attraction was home ownership, enjoyed by only 10% of the British population in 1914 but 31% in 1939.[48] Plotlands meant owning a place of one's own, like, 'a panacea ... to cure all afflictions.'[49] Huts and chalets proliferated on England's south coast from the 1890s onwards, beside the more expensive bungalows of retired colonial officers (using a Bengali building type). A plot, then, was, 'a poor person's paradise, with a choice of everything from a cheap, ready-made seaside bungalow to a bare strip of shingle on which at weekends you could build your own shelter.'[50]

But some plotlands were objects of capitalist exploitation. At Peacehaven, entrepreneur Charles Neville bought 415 acres at £15 per acre in 1914, adding a further 150 acres four years later. He then sold plots at £50 and £100 through a competition for the site's name. Legal complications limited his scheme, so that instead of founding a small town by the sea as intended, plots were built on haphazardly. Hardy and Ward remark,

> The result of all this was that Peacehaven soon acquired a classic plotland landscape of scattered development, with vacant plots and waste-land interspersed between completed units. A second plotland characteristic was the visible evidence of what was, perhaps, an understandable reluctance on the part of the local authority to invest in Peacehaven's straggling landscape.[51]

The first chalet was built in 1921. Prefabricated homes were also available. Plotlanders were ex-servicemen using a service gratuity for the purchase; and retired shopkeepers or artisans. Hardy and Ward note that apart from having a sea view a plot was, 'a mark of cultural respectability and pedigree for the upstart community.'[52] It was an upper-working- and lower-middle-class society within society, a little further down the social pyramid than aspirant suburbanites.

Plotlanders, like ramblers, reclaimed the land as a means to reclaim social organisation; and embellished their chalets in various personal ways. They lived as they chose. This is a-social in a way, but asserts a right to belong which is a social reaction against a state which fails to meet the needs of non-privileged classes, leading to a claim to an unregulated life. That this appeals to a pre-industrial ethos illustrates the general cultural condition of the time, being modern in some ways, regressive in others, and throughout a rejection of cities as where a life might be lived to the full. The difficulty is that while most people's lives were not lived to the full in cities—due to pollution, bad housing and restricted access to education, lack of health care, and not enough money—the source of these woes was not the modern city but the economic system which made its experiences so unequally available.

NOTES

1. *Building News* (1899) vol. 76, p. 60, quoted, Hubbard, E. and Shippobottom, M., *A Guide to Port Sunlight*, Liverpool, Liverpool University Press, p. 41 [no original author/title given].
2. Lever, W.H. [1907] (2005) *Visit of International Housing Conference to Port Sunlight*, pamphlet, pp. 10–11, quoted, Hubbard and Shippobottom, *Guide to Port Sunlight*, pp. 19, 24.
3. Hubbard and Shippobottom, *Guide to Port Sunlight*, p. 51.

4. Tennyson, A. [1835–42] (n.d. c. 1900) *The Poetical Works*, intro. Waugh, A., London, Collins, p. 141.
5. Bournville Estate, (n.d. c. 1945) *Sixty Years of Planning: The Bournville Experiment*, Bournville, Bournville Estate, pp. 24–25.
6. Jolly, W.P. (1976) *Lord Leverhulme, a Biography*, London, Constable, pp. 80–81.
7. Liversidge, M., J. H. (1997) 'Virgil in Art,' Martindale, C., ed., *Cambridge Companion to Virgil*, Cambridge, Cambridge University Press, p. 99.
8. Nicholson, A.,2(008) *Arcadia: The Dream of Perfection in Renaissance England*, London, Harper, p. 1.
9. Duncan-Jones, K. (2008) Introduction, Sidney, P., [1580] *The Old Arcadia*, Oxford, Oxford University Press, p. xiv.
10. Duncan-Jones, *The Old Arcadia*, p. xiii.
11. Sidney, *The Old Arcadia*, p. 4.
12. Sidney, *The Old Arcadia*, p. 4.
13. Sidney, *The Old Arcadia*, p. 11.
14. Nicholson, *Arcadia*, p. 1.
15. Nicholson, *Arcadia*, p. 14.
16. Nicholson, *Arcadia*, p. 36.
17. Nicholson, *Arcadia*, p. 68.
18. Hill, J. (2016) *A Landscape of Architecture, History and Fiction*, London, Routledge, p. 45.
19. Coke, T.W., quoted in Porter, R. (1991) *English Society in the Eighteenth Century*, London, Penguin, p. 60, cited in Hill, *A Landscape of Architecture*, p. 80 [no original source given].
20. Illustrated, Hill, *A Landscape of Architecture*, pp. 78–79.
21. Hill, *A Landscape of Architecture*, p. 61.
22. Colonna, F. [1499] (1999) *Hypnerotmachia Poliphili, The Strife of Love in a Dreame*, trans. Godwin, J., London, Thames and Hudson.
23. Daniels, S. (1993) *Fields of Vision: Landscape Imagery and National Identity in England and the United States*, Cambridge, Polity, p. 80.
24. Daniels, *Fields of Vision*, p. 82.
25. Daniels, *Fields of Vision*, p. 98.
26. Pugh, S. (1998) *Garden, Nature, Language*, Manchester, Manchester University Press, p. 35.
27. Woodward, C. (2002) *In Ruins*, London, Vintage, p. 121.
28. Pugh, *Garden, Nature, Language*, pp. 21–22.
29. Pugh, *Garden, Nature, Language*, p. 57.
30. Darley, G. [1975] (2007) *Villages of Vision: A Study of Strange Utopias*, Nottingham, Five Leaves, p. 15.
31. Darley, *Villages of Vision*, p. 17.
32. Darley, *Villages of Vision*, p. 27 [no source given].

33. Hall, P. and Ward, C. (1998) *Sociable Cities: The Legacy of Ebenezer Howard*, Chichester, Wiley.
34. Howard, E., [c. 1900] unpublished article for Contemporary Review, quoted in Hall and Ward, *Sociable Cities*, p. 21, citing Beevers, R. (1988) *The Garden City Utopia*, London, Macmillan, p. 80.
35. Illustrated, Hall and Ward, *Sociable Cities*, p. 18.
36. Howard, E. (1898) *Tomorrow! A Beautiful Path to Real Reform*, London, Swan Sonnenschein, p. 5, quoted in Hall and Ward, *Sociable Cities*, p. 18.
37. Judge, T. (2014) *Gardens of Eden: British Socialism in the Open Air 1890–1939*, London, Alpha House, p. 63.
38. Sloterdijk, P. [2005] (2013) *In the World Interior of Capital*, Cambridge, Polity, p. 170.
39. Sloterdijk, *In the World Interior of Capital*, p. 171.
40. Darley, *Villages of Vision*, p. 143.
41. Proceedings of the Countryside and Footpaths Preservation Conference 1928, quoted in Matless, D. (1998) *Landscape and Englishness*, London, Reaktion, p. 54; 294, n. 140.
42. Matless, *Landscape and Englishness*, p. 54.
43. Judge, *Gardens of Eden*, p. 93.
44. Illustrated, Ross, C. with Bennett, O. (2015) *Designing Utopia: John Hargrave and the Kibbo Kift*, London, Philip Wilson, p. 40.
45. Ross and Bennett, *Designing Utopia*, p. 43.
46. Judge, *Gardens of Eden*, p. 45.
47. Hardy, D. and Ward, C. (1984) *Arcadia for All: The Legacy of a Makeshift Landscape*, Nottingham, Five Leaves, pp. 11–12.
48. Hardy and Ward, *Arcadia for All*, p. 17.
49. Hardy and Ward, *Arcadia for All*, p. 25.
50. Hardy and Ward, *Arcadia for All*, p. 60.
51. Hardy and Ward, *Arcadia for All*, p. 76.
52. Hardy and Ward, *Arcadia for All*, p. 83.

CHAPTER 3

Drawing a Line

Abstract This chapter reconsiders René Descartes' *Discourse* (1637), and the architectural metaphor he uses to indicate the certainty of an abstract system. The *Discourse* can be read as foundational to modern thought, and its architectural metaphor as foundational for the separation of an abstract process of design from the actualities of occupation in modern planning and architecture. The context for Descartes' text is the Thirty Years' War, from which his evasion of doubt is a retreat. Henri Lefebvre is critical of Cartesian space, seeing it as the dominant element within a society's spatial practices; but, for Lefebvre, the conceived space of plans is always in a dialectic relation to the lived spaces of occupation.

Keywords Descartes • Regular spaces • Lefebvre • Conceived space • Lived spaces

The previous chapter looked at the development of an Arcadian current in British culture, from the sixteenth century onwards. In this chapter I move to European thought in the seventeenth century, and René Descartes *Discourse on Method* (1637). Descartes uses an architectural metaphor—an engineer draws regular spaces on a blank ground—which, to me, sums up modernity's dominant spatial practice: the abstract space of plans constituting a seamless dimension. In another sphere, material adaptations and emotional ownership denote contingent space. Abstract space reduces the

© The Author(s) 2021
M. Miles, *Paradoxical Urbanism*,
https://doi.org/10.1007/978-981-15-6341-6_3

world to cartographic representation using regular coordinates; against this, the spaces of occupation and everyday life reassert. To use a phrase from French Marxist sociologist Henri Lefebvre, a right to the city. I begin by summarising what Descartes said and the context in which he said it (the Thirty Years' War), then move to interpretations of his architectural metaphor. I juxtapose this to the emphasis in Lefebvre's critique of Cartesianism on mutability and the always incomplete processes of spatial practice, which informs my discussion of modern urbanism in Chap. 4.

METHODS OF THINKING

Descartes' text was first published, some years after he wrote it, as *Discourse on the Method of Rightly Conducting One's Reason and Seeking the Truth in the Sciences, and in Addition the Optics, the Meteorology and the Geometry, which are essays in the Method.*[1] I regard it as foundational to the modern idea of space as making new, drawing on a blank ground; and as a precursor to the separation in modern architecture and planning of design from material production.

Descartes conducts his thinking away from the world, in an interior space. There, in a stove heated room, he authors a text which refuses representations of knowledge in books, the experiential knowledge of travel, and the tacit knowledge of bodily sensation, all of which are contingent on an exterior realm and might thus be unreal. The borders of the interior and exterior realms are porous, but the outcome is a dualism, sometimes called the mind–body split—an unhelpful phrase—but which is, more accurately, an attempt to preserve religious orthodoxy in terms of the soul (which leaves the body), and to state the subject's separation from the objects of perception other than in observing its own mental operations.

Descartes' philosophical quest is haunted by doubt. His aim is to discover a certainty beyond doubt, or, philosopher John Cottingham writes, 'to see whether there is anything at all that survives the doubt.'[2] Descartes develops this rationality beyond doubt in optics, geometry, and mathematics, all closed systems; but he begins the *Discourse* by asserting, 'I have never presumed to regard my mind as in any way more perfect than the average.'[3] He sees himself as fortunate in having acquired a step-by-step mode of enquiry, but admits that, 'I may be mistaken, and what I took for gold and diamonds may be no more than a little brass, and some fragments of broken glass.'[4] That could be a metaphor for modernity, like

T. S. Eliot's line in *The Waste Land* (1922), 'These fragments I have shored against my ruins.'[5]

Descartes' method is, at least, paradoxical, both an affirmation of the possibility of certainty and an acceptance of fragility in his attempt (or essay, using a term from Montaigne). He sets out the areas of learning he has undertaken: ancient and modern languages, poetry, and mathematics, '[which latter] gave me the most pleasure because of the certitude and evidential character of its reasonings.'[6] In comparison to mathematics, the 'moral treatises of the ancients' are, 'splendid palaces built on sand and mud.'[7] But Descartes' own situation is nearly as insecure: 'I found myself hampered by so many doubts and errors that the only benefit of my efforts ... [was] the increasing discovery of my own ignorance.'[8] Conventional means to knowledge—books, travel, and sensory experience—are open to error or illusion; but at this point he, significantly perhaps, mentions his withdrawal into a solitary space for contemplation, in a passage in the *Discourse* which I quote at length because it is pivotal,

> I was at one time in Germany, attracted thither by the wars which are not yet ended, and was on my way from the coronation of the Emperor to join the army, when winter brought me to a halt in quarters where, with no society to distract me, and no cares or passions to disturb me, I spent the day in a stove-heated room, with all the leisure in the world to occupy myself with my own thoughts. Among these, one of the first that came to my mind was that there is often less perfection in what has been put together bit by bit, and by different masters, than in the work of a single hand. Thus we see how a building, the construction of which has been undertaken and completed by a single architect, is usually superior in beauty and regularity to those that many have tried to restore by making use of old walls which had been built for other purposes. So, too, those old places which, beginning as villages, have developed in the course of time into great towns, are generally so ill-proportioned in comparison with those an engineer can design at will in an orderly fashion ... Finally, I reflected that, as we have all been children, long governed by our appetites and tutors ... it is almost impossible that our judgements should be as clear and as well-founded as they would have been if we had had the use of our reason from birth and had never been governed by anything else.[9]

This is pivotal because the closed, abstract systems he develops are, in effect, mirrors of this enclosed space into which exterior reality does not intrude.

Descartes' architectural metaphor of a building designed in a single process is essentially modern. He was probably aware of early planned cities such as Charleville (1605) or Nancy (1611); but he looks to a method of reasoning, not building, its analogy to a design being a way to propose a system of knowledge based on a single principle. In keeping, a realisation that the thinking self exists—*cogito ergo sum*—alone ends his haunting by doubt of the real.

Descartes' method has four rules; to accept nothing as true which can be doubted; to divide each problem into parts; to think in an orderly way, moving from the simple to the complex; and to review all steps in an argument, so that, 'I could be sure I had missed nothing.'[10] The knowledge gained in geometry—a theorem once proven is proven forever—is a complete, closed system; mathematics similarly relies on the consistency with which rules are applied. $2 + 2 = 17$ is not viable in philosophy or ordinary life. And the authorship of thought relies on an isolation of the mind which produces it, unalloyed by the vicissitudes of the world and its troubles. Descartes writes, moving to the Netherlands in 1629, 'amid a multitude of busy people, more concerned with their own business than curious about the affairs of others ... I have been able to live as solitary and undisturbed as in the most remote of deserts.'[11]

Descartes' discussion of mind and body is slightly ambivalent. Cottingham cites Descartes' letter to Princess Elizabeth of Bohemia: '[He] explores the paradox that while philosophical reason teaches us that mind and body are distinct, our everyday human experience shows us they are united.'[12] Mind and body are separate, the mind as the location of thought, the body in the material world of shifting appearances. But Descartes presents this as a question as to whether the mind (or soul) exists separately from the body, as said, to avoid religious controversy. Critical theorist Andrew Benjamin cites Descartes' *Meditations* (1641) to relate the mind–body problem to that of knowledge, which requires that what is known is tested by proof, beyond perception or belief. Benjamin summarises from the *Meditations*, 'What are at stake ... are two related projected movements. The first is formulating a new set of criteria ... the second is that this formulation must take place anew.'[13] Mind is the location of scrutiny, leading to Cartesian dualism in that what is scrutinised is constructed as outside the scrutinising self: a subject-object dualism. Benjamin continues that Cartesian dualism demands a separation of mind and body to enable, 'the centrality and supremacy of the mind and the subsequent reintroduction of the body,' based on that separation.'[14] He adds, 'The removal of the body is necessary in order that the primacy of the ego ... can emerge. It is thus that the body can come to be represented.'[15]

This introduces a difficulty: modern science sets aside precedent and previously held beliefs but constructs, Benjamin says, 'a present in which nothing can be added' so that judgement is necessarily repeated; yet this, 'becomes impossible because of the eventual impossibility of sustaining *forgetting* within it.'[16] Myth is abolished, but its abolition is reiterated as a new myth, I might say, borrowing an idea from Theodor Adorno and Max Horkheimer.[17]

But more important, perhaps, is the centrality which Descartes lends to authorship—*I* think therefore *I* am—a product of self-imposed exile. This is where the separation from the body in which the self thinks occurs: 'I had in my mind the ideas of many sensible and corporeal objects; I could suppose that I was dreaming ... but ... could not deny that those *ideas* were really in my mind.'[18] Ideas are real. The reality of what they represent has no such certainty; and this is realised in the mind of a subject (self) aware of objects the imagination of which, more than perception, instantiates the existence of the subject.

The privileging of the self—*ce moi* in Descartes' French text—separates it from society in a subject-object dualism. Today, after the reputed death of the author in post-structuralism, an assertion of authorship (or authority) would be deconstructed, and attention drawn to the intersection of interior and exterior realms to the point of non-differentiation (just as the dualism of cause and effect dissolves when effects become causes). In the 1630s–1640s the interiority of the thinking subject was a radical departure from a system inherited from religious instruction based on repeated re-readings of scripture. But the idea of a subject is, despite its claimed autonomy, a response to doubt, and underneath to mortality. So, I think Descartes' thought is foundational for modernity in the imagination of regular systems, but at the cost of setting aside the realms in which mortality is intimated, such as the body and its sensations (or dreams).

Social theorist Peter Wagner alludes to such an uncertainty when he argues that Descartes' rationalism rests on, 'a radical positing of subjectivity'[19] in a search for certainty distancing the observed world from the observer to find the indisputable among the falsehoods. But this, he says, is a product of a uncertainty requiring an overarching set of rules for society: 'The situation of uncertainty that was experientially self-evident in a situation of devastating religious and political strife had to be overcome by appeal to an instance that ... could only be outside such experience.'[20] Devastating strife, then, in which mortality was all too real and sudden, is the context for the *Discourse*.

COSMOPOLIS?

As said, the *Discourse* was written and published during the Thirty Years' War (1618–1648), a conflict between Catholic and Protestant dynasties and ideologies in which Descartes was a gentleman observer at the Emperor's court. That is why he travelled in Germany and found himself in a stove-heated room. Historian Stephen Toulmin writes of this period that, 'rival militias and military forces consisting largely of mercenaries fought to and fro, again and again, over the same disputed territories ... in the name of theological doctrines that no one could give any conclusive reasons for accepting.'[21] Around a third of the population of the land which now constitutes Germany and the Czech Republic were killed, either directly in fighting or through famines resulting from the destruction of crops, or in the displacement following the burning of villages and towns. Both sides committed atrocities. Toulmin asks, 'In this blood-drenched situation, what could good intellectuals do?'[22] They could maintain Renaissance humanism, or withdraw. Or,

> Might not philosophers discover ... a new and more rational basis for establishing a framework of concepts and beliefs capable of achieving the agreed certainty that the skeptics had said was impossible? If uncertainty, ambiguity, and the acceptance of pluralism led, in practice, only to an intensification of the religious war, the time had come to discover some rational method for demonstrating the essential correctness or incorrectness of philosophical, scientific, or theological doctrines.[23]

This is the reason for the *Discourse*. But Toulmin identifies another response in the idea of Cosmopolis, a fusion of two systems: that of *cosmos*, the natural world and natural sciences, with *polis*, political and social organisation. Cosmopolis unites these ideas in a single ideal as a means to resolve the separation of the natural from the social and political world.

Toulmin argues that Cosmopolis figures in the rise of nation-states in the seventeenth and eighteenth centuries, replacing dynastic feudalism;

and in the rise of natural science in the same period. Both seek stability, and the nation-state constructs it in the relations of states to each other while also constructing a hierarchy in their internal social structures, making modern political institutions on a seemingly non-contingent basis:

> It was important to believe that the principles of stability and hierarchy were found in all of the Divine plan, down from the astronomical cosmos to the individual family. Behind the inertness of matter, they saw in Nature, as in Society, that the actions of lower things depended on, and were subordinate to oversight and command by higher creatures ... The more confident one was about subordination and authority in Nature, the less anxious one need accordingly be about social inequalities.[24]

That is a reading of *cosmos* and *polis* in seventeenth-century history; another, longer, view of *polis* is that it is both the built city *and* its social organisation, cited by architect Krzysztof Nawratek in a comparison of the Greek *polis* and Latin *urbs*: '*Polis* is both understood as a physical city and the community living in it, while the Roman model requires differentiation between the material and spatial *Urbs*, and *Civitas* as a community of residents.'[25] I return to this in Chap. 6. Here, I end this part of the chapter by quoting Toulmin again:

> The comprehensive system of ideas about nature and humanity that formed the scaffolding of Modernity was thus a social and political, as well as a scientific, device: it was seen as conferring Divine legitimacy on the political order of the sovereign nation-state. In this respect, the world view of modern science ... won public support around 1700 for the legitimacy it apparently gave to the political system of nation-states as much as for its power to explain the motions of planets, or the rise and fall of tides.[26]

Cosmopolis, like Descartes' subjective rationality, offers freedom from conflict at the price of subordination to a seamless, universal ordering. The abstract ideal of Cosmopolis appears to be a parallel response to the same set of historical conditions which shaped Descartes' solitary thoughts, the two models of apprehending reality shedding light on each other as idealistic and fanciful, yet desperately seeking a basis for living which will not shatter.

DRAWING A LINE

Descartes sits alone without the distractions of society or the disturbances of passions. He imagines a unity of design as preferable to, more ordered than, an accumulation of forms. He contrasts a village which has grown into a town over time with a city, 'which an engineer can design at will in an orderly fashion.'[27] Then he adds a colonial analogy:

> In the same way I fancied half-savage nations, who had gradually become civilised, but who had made their laws by degrees as the need arose to counter the harm done by crimes and disputes, could never be as well-regulated as those who, from the beginning of their associations, had observed the decrees of some pungent lawgiver.[28]

The law-giver is god, providing an oblique authority to the idea expressed earlier in the text that a single design for a city is better than the mess of accumulation and use. This leads to the image of an engineer drawing a line on a blank ground, to build a body of knowledge by erasing inherited systems, creating a *tabula rasa*. I want to look again at the image of the engineer, citing literary historian Claudia Brodsky Lacour, who uses her own translation of the *Discourse* as well as a more recent English version than that I use above.[29]

Instead of, 'which an engineer can design at will in an orderly fashion'[30] (from my dog-eared Penguin edition), Lacour writes, 'that an engineer traces on a vacant plain according to his free imaginings [or fancy].'[31] The free imagining is Descartes' act of thinking, constructing a new method to know the world: 'The image of a single architect drawing a comprehensive plan on a "vacant plain" is the "thought" Descartes opposes both to travel and to reading.'[32]

Descartes states his intention as building a method for his own use, not initiating a universal reform. But Lacour argues that, unlike the engineer drawing freely, Descartes discourses in a way which is, 'both representational and intentional.'[33] She differentiates two meanings of the word design, as plan (*dessin*) and as intention (*dessein*):

> While Descartes intends to order his own thinking; in doing so, he imagines the engineer designing space. Descartes commonly uses *dessein* in the *Discours* and elsewhere when stating his speculative plan or intention, he first uses *dessin*, an architect's plan, when, after presenting the four rules of procedure, he describes what one must do "to rebuild a house" in addition

to "having carefully traced its ground plan" (*dessin*). ... [Here] Descartes develops and significantly alters the image of the act of architectural drawing he had "thought" of in his stove-heated room.[34]

This might seem obscure philology but I don't think it is: the intention Descartes has for his thinking is expressed undemonstratively as a modest act for his own work; the plan (*dessin*) of a city or a built space is, in contrast, instrumental. Although Descartes uses the image of an engineer freely drawing a line *as a metaphor*, the image, like the imagined regular space, is an act of inscription on a blank sheet as if nothing was there. For Lacour, this, 'non-figural delineation,' an image with no mimetic aspect, is fundamentally discursive.[35] She continues,

> The act of architectural drawing that Descartes describes is the outlining of a form that was not one before. That form would combine reason ... with imaginative freedom ... [it] intervenes in a space where nothing was, on a surface ... where nothing else is. The order of its "places regulières" is the image of imagination engineering a method that is free of historical and intellectual as well as physical constraints.[36]

The line is fanciful yet purposeful: the engineer draws it because it can be transposed to the demarcation of real spaces. Modern architecture takes this as a norm; the non-descriptive, however, becomes prescriptive while intention (*dessein*) becomes the basis for a city's or a building's design when power, money and technology allow it. Freely thought, the line is a sign which carries its own implicit power. The knowledge of architectural drawing becomes power-over reality, making a new world in a seamless space of regular coordinates where it is unchallenged by history or personality.

It is not surprising that this vein of rationalism runs parallel to a growth in map-making in Europe, from the initial manufacture of a geographer's globe in 1492 to two-dimensional projections of the round Earth in maps used in navigation, and to maps of cities surveyed accurately for taxation. Geographer John Pickles writes of the central role of maps in the establishment of nation-states:

> The map ... has been an essential tool in territorialising the state by extending systems of policing and administration, and in establishing a sense of national identity at home and abroad (sometimes in the face of explicit internal disunity or rebellion). The state must consistently attempt to capture the

discursive and ideological field not only through the more obvious organs of public opinion, but also by the appropriation of space (and the map) to its purposes and by the symbolic constitution of mapped space as national space.[37]

Maps exert power-over. They can be used for this end because they re-codify multiple and complex realities, subject to claim and counter-claim, as unified, coherent, irrefutable fact. It is this power relation in Cartesian space which leads to another power relation, that of the centre and the margin, (which was not marginal before this inscription of power-over).

Geographer Doreen Massey argues that received ideas, including those of space, become ingrained: 'We have inherited an imagination so deeply ingrained that it is often not actively thought. Based on assumptions no longer recognised as such, it is an imagination with the implacable force of the patently obvious. That is the trouble.'[38] An example is, precisely, the dualism of appearances: light-dark, good-bad, me-you, on-off, or the world as a light switch (before invention of the dimmer). To question dualism is thus a point of departure for post-Cartesian critical thought, leading geographer Edward Soja to write, from Lefebvre, 'thinking trial-lectically' enables a radical understanding of the, 'limitless composition of lifeworlds.'[39]

NUANCED SPACES

In *The Production of Space*, Lefebvre comments that Descartes shifted phi-losophy from the rehearsal of 'thought thought' to the practice of 'think-ing thought,' and from the objects of thought to, 'the operation of knowing.'[40] He ascribes a development from Descartes to Karl Marx, moving from 'works generated by knowledge,' to, 'things in industrial practice.'[41] This is the philosophy of practice, the dialectical materialism which fuses the Reason of German Idealism with the Materialism of Ludwig Feuerbach's idea that environments shape individuals. That is, the conditions inhabited by a subject condition that subject's mentality, but in dialectical materialism those conditions are re-conditioned by the sub-ject's agency.

In terms of built environments, the question is how they are *produced*, not given. But that is my gloss. Lefebvre writes, 'What we are concerned with … is the long *history of space*, even though space is neither a subject nor an object but rather a social reality.'[42] This history can encompass both representations of space and representational spaces. The two terms are

confusingly alike: one is plural–singular and the other is singular–plural.
Representations of space include drawings and plans; representational
spaces are the endlessly adapted sites of occupation. In representations of
space, Descartes' *dessin* is the instrument of *dessein*. That is, an abstract
line delineates a space to be built; the intention is separated from the work
of material production so that the intellectual process of design is seen as
superior to building. Architects do not generally get their hands dirty, let
alone—with exceptions—co-design spaces with users. Lefebvre says, 'the
producers of space have always acted in accordance with a representation,
while the users passively experienced whatever was imposed upon them…
inserted into, or justified by, their representational space.'[43]

Reflecting on modernist architecture and planning, and citing Le
Corbusier (who did both), and Haussmann's remodelling of Paris in
the1850s–1860s, Lefebvre asserts, 'The outcome has been an authoritarian and brutal spatial practice … what is involved in all cases is the effective
application of the analytic spirit in and through dispersion, diversion and
segregation.'[44] This produces a reductive or strategic space:

> The ruling classes seize hold of abstract space as it comes into being (their
> political action occasions the establishment of abstract space, but it is not
> synonymous with it); and they then use that space as a tool of power, with-
> out for all that forgetting its other uses: the organisation of production and
> of the means of production – in a word, the generation of profit.[45]

Then, Cartesian rationality produces spatial distinctions such as zoning—the demarcation of distinct spaces for, for instances, administration,
industry, dwelling or leisure—which break up space by function, frag-
menting a city under a supposedly (but far from) neutral planning regime.
In this technocracy, classical reason degenerates into technological rationality; or, as Lefebvre adds, it is transformed, 'into the absurdity of pul-
verised reality.'[46] A case of this is architectural drawing:

> Within the spatial practice of modern society, the architect ensconces him-
> self [sic] in his own space. He has a representation of this space, one which
> is bound to graphic elements – to sheets of paper, plans, elevations, sections,
> perspective views of façades, modules, and so on. This *conceived* space is
> thought by those who make use of it to be *true*, despite the fact – or because
> of the fact – that it is geometrical: because it is a medium for objects, an
> object itself, and a locus of the objectification of plans.[47]

Lefebvre argues that representations of space and representational spaces are co-present in an antagonistic relation within a society's spatial practices. But the relation is not equal. It is, I could say, power-over against power-to, instrumentalism against empowerment; or, using a simpler formulation. That Lefebvre's conceived (planned) space dominates lived spaces.

Lefebvre insists on reading Descartes in context of the conditions in which he thought, too, adding a criticality inadmissible in a closed system. Coincidentally, this is the basis as well for the Frankfurt School's critical theory: it is critical of the conditions of its production. Political geographer Stuart Elden elaborates that, for Lefebvre, Descartes' thinking is understood in, 'a range of extremely complicated and contradictory ways.'[48] These include the subjectivity in which the idea of objective knowledge is imagined, I suggest, while the implicit power of representations of space tends always to be blurred and interrupted by occupation and use.

Lefebvre reads Descartes' rationality as arising when the urban becomes the dominant form of society. Elden summarises this as, 'urbanity (cultivated) is opposed to rusticity (naïve and brutal).'[49] With industrialisation, the process moves further into instrumental space as urban society recodes the countryside as well as cities as zones of spatial and social reordering. But power leaks. Representational spaces are spontaneously reclaimed. For Elden, the social relations produced in industrialisation are one side of the dialectic; the other side is that lived spaces can be creatively retrieved from the seamlessness of planned space: 'Lefebvre suggests that the city can be a work in the sense of a work of art, because it is not simply organised and instituted, but can also be modelled and appropriated.[50]

I am cautious about cities being works of art because this aestheticises realities which are social, economic and political as well as cultural, which change, while a finished work of art remains the same except inasmuch as a viewer completes it through personal interpretation or association. Leaving that aside, I think Elden has in mind the imaginative processes by which spaces are re-appropriated. This is the basis of Lefebvre's idea of a right to the city. In *The Right to The City* (1968), he writes that a supposed right to nature has evolved in leisure pursuits involving the countryside— 'against noise, fatigue, the concentrationary universe of cities'—so that nature, 'enters into exchange value and commodities, to be bought and sold.'[51] Against this pseudo-right, the right to the city is, 'like a cry and a demand … [which] slowly meanders through the surprising detours of nostalgia and tourism, the return of the heart of the traditional city, and the call of existent or recently developed centralities.'[52] This right is not a permit, like a visiting right, but a right to urban life understood, of which he says only the working class will be the agents. This was published in 1968 as students occupied the universities of the Sorbonne and Nanterre, where Lefebvre taught sociology. Although, then, Lefebvre revises Marxism he retains its class structure while others there at the time, such as Herbert Marcuse, see a different driving force in radical social change, and a cultural as well as social impetus.[53] I return to this in Chap. 6.

For Lefebvre, to see space as an object is reductive but contestable: 'abstract space ... makes the relationship between repetition and difference a more antagonistic one.'[54] Just like the body, that is, society is not an abstract entity, and cannot live without producing differences which fracture the seamless space of drawing a line. Lefebvre continues that the architect occupies an uncomfortable position:

> As a scientist and technician, obliged to produce within a specified framework, he [sic] has to depend on repetition. In his search for inspiration as an artist, and as someone sensitive to the use and to the user, however, he has a stake in difference. He is located willy-nilly within this painful contradiction, forever being shuttled from one of its poles to the other.[55]

This echoes the sensuous human activity of Marx' *Theses on Feuerbach* (1845): 'Feuerbach ... appeals to sensuous contemplation; but he does not conceive sensuousness as practical, human-sensuous activity.'[56] Sensuousness is displaced by representations of space, but to displace is not to eliminate; and the reductive aspect of Cartesian space can be ruptured to enable a reclamation of the open-ended multiplicity of association and occupation. This is in effect a social revolution. Architect Luksz Stanek writes, 'the reductionism of the Cartesian system of representation (the very cause of its practical success) ... endowed space with a simultaneous tendency towards homogenisation and fragmentation.'[57]

These tendencies are, Lefebvre argues, interdependent,[58] and, 'paradoxically, united yet disunited, joined yet detached from one another.'[59] It is the sameness and seamlessness of Cartesian space which ensures its fracture in the limitations of representation according to a closed system of coordinates, and the inevitability that the homogeneity in question will at some point collapse or require maintenance by force or the power of money. The inherent failure of Cartesianism is that its denial of the material is prone to leaks, and the production of unstable effects. For instance, the inscription of representations of space on a modern city tends to produce fragmentation by function, or single-use zoning. This, in turn, leads to a dominance of designation over experience—when people who live in cities move around them and have multiple, overlapping social and communicative networks—for which an analogy might be a dominance of rules over everyday usages in verbal language. Lefebvre insists that both elements are always present, in contention.

The failure of Cartesian space can, then, be said to be a failure of binary divides, as Soja says, describing Lefebvre's theories as a, 'nomadic meta-Marxism.' He continues,

> For Lefebvre, reductionism … begins with the lure of binarism, the compacting of meaning into a closed either/or opposition … Whenever faced with such binarised categories … Lefebvre persistently sought to crack them open by introducing an-Other term, a third possibility of moment that partakes of the original pairing but is not just a simple combination or an in-between position along some all-inclusive continuum.[60]

The point is that conceived space and lived spaces are co-present (as said above) in a society's spatial practices, and are engaged in dispute. Why should this matter in a book on urbanism? Because it offers a method, more open than Cartesianism, for a critical approach to how cities are designed, constructed, and occupied. Lefebvre was aware of the planning and design of new towns in France in the post-war period, and critical of the rigid spatial demarcations involved. His writing can be as abstract as Descartes' at times, if more musical—*The Production of Space* is like a fugue, theme and variations reiterated—but signals an allegiance to a rehabilitation of dwelling (the life of users) and, by implication, a revolution in spatial practice which reinstates dwellers as legitimate producers of a city.

NOTES

1. Chavez-Arvizo, E., ed. (1997) *Descartes: Key Philosophical Writings*, Ware (Herts), Wordsworth Editions, p. xiii.
2. Cottingham, J., (1997) *Descartes*, London, Phoenix, p. 20.
3. Descartes, R., [1637] (1960) *Discourse on Method*, trans. Wollaston, A., Harmondsworth, Penguin, p. 36.
4. Descartes, *Discourse on Method*, p. 37.
5. Eliot, T. S. [1922] (1971) *The Waste Land*, facsimile and transcript, London, Faber and Faber, p. 146 (line 430).

6. Descartes, *Discourse on Method*, p. 41.
7. Descartes, *Discourse on Method*, p. 41.
8. Descartes, *Discourse on Method*, p. 39.
9. Descartes, *Discourse on Method*, pp. 44–46.
10. Descartes, *Discourse on Method*, p. 50.
11. Descartes, *Discourse on Method*, p. 60.
12. Cottingham, *Descartes*, p. 16.
13. Cottingham, *Descartes*, p. 16.
14. Benjamin, A., (1993) *The Plural Event: Descartes, Hegel, Heidegger*, London, Routledge, p. 46.
15. Benjamin, *The Plural Event*, p. 47.
16. Benjamin, *The Plural Event*, p. 60 (my italics).
17. Adorno, T.W. and Hormheimer, M. [1944] 1997 *Dialectic of Enlightenment*, London, Verso, pp. 43–80.
18. Descartes, *Discourse on Method*, p. 63 (my italics).
19. Wagner, P., (2001) *Theorising Modernity*, London, Sage, p. 17.
20. Wagner, *Theorising Modernity*, p. 19.
21. Toulmin, S. (1990) *Cosmopolis: The Hidden Agenda of Modernity*, Chicago, University of Chicago Press, p. 54.
22. Toulmin, *Cosmopolis*, p. 55.
23. Toulmin, *Cosmopolis*, p. 55.
24. Toulmin, *Cosmopolis*, p. 128.
25. *Nawratek, K. (2019) Total Urban Mobilisation: Ernst Jünger and the Post-Capitalist City*, Singapore, Palgrave, p. 57.
26. Toulmin, *Cosmopolis*, p. 128.
27. Descartes, Discourse, pp. 44–45.
28. Descartes, *Discourse*, p. 45.
29. Descartes, R. [1637] (1925) *Discourse de la méthode*, ed. Gilson, E., Paris, Virin; (1980) *Discourse on method*, trans. Cress, D.A., Indianapolis, Hackett.
30. Descartes, *Discourse*, p. 44.
31. Lacour, C.B. (1996) *Lines of Thought: Discourse, Architectonics, and the Origin of Modern Philosophy*, Durham (NC), Duke University Press, p. 33.
32. Lacour, *Lines of thought*, p. 33.
33. Lacour, *Lines of Thought*, p. 35.
34. Lacour, *Lines of Thought*, p. 36.
35. Lacour, *Lines of Thought*, p. 36.
36. Lacour, *Lines of Thought*, p. 37.
37. Pickles, J., (2004) *A History of Spaces: Cartographic Reason, Mapping and the Geo-coded World*, London, Routledge, p. 39.
38. Massey, D., (2005) *For Space*, London, Sage, p. 17.

39. Soja, E. (1996) *Thirdspace: Journeys to Los Angeles and other Real-Imagined Places*, Oxford, Blackwell, p. 70.
40. Lefebvre, H., [1974] (1991) *The Production of Space*, Oxford, Blackwell, 1991, pp. 114–115.
41. Lefebvre, *The Production of Space*, p. 115.
42. Lefebvre, *The Production of Space*, p. 116 (italics original).
43. Lefebvre, *The Production of Space*, pp. 43–44.
44. Lefebvre, *The Production of Space*, p. 308.
45. Lefebvre, *The Production of Space*, p. 314.
46. Lefebvre, *The Production of Space*, p. 317.
47. Lefebvre, *The Production of Space*, p. 361.
48. Elden, S. (2004) *Understanding Henri Lefebvre*, London, Continuum, p. 90.
49. Elden, *Understanding Henri Lefebvre*, p. 131.
50. Elden, *Understanding Henri Lefebvre*, p. 134, citing Lefebvre, H. (1972) *Espace et politique: Le droit de la ville*, Paris, Anthropos, p. 74.
51. Lefebvre, *The Right to the City* [1968] (1996), in *Writings on Cities*, ed. Kofman, E and Lebas, E., Oxford, Blackwell, pp. 157–158.
52. Lefebvre, *The Right to the City*, p. 158.
53. Marcuse, H., (1969) *An essay on Liberation*, Harmondsworth, Penguin.
54. Lefebvre, *The Production of Space*, p. 396.
55. ibid.
56. Marx, K., [1845] (1968) *Theses of Feuerbach*, V, in *Marx and Engels, Selected Works*, London, Lawtence and Wishart, p. 29.
57. Stanek, L. (2008) 'Space as concrete abstraction,' Goonewardena, K., Kipfer, S., Milgrom, R. and Schmid, C., eds., *Space, Difference, Everyday Life: Reading Henri Lefebvre*, London, Routledge, p. 71.
58. Lefebvre, *The Production of Space*, pp. 97–98; 287; 308.
59. Stanek, 'Space as concrete abstraction,' p. 71.
60. Soja, *Thirdspace*, p. 60.

CHAPTER 4

The Contradictions of Modernism

Abstract This chapter extends the critical analysis of space in Chap. 3 by examining specific sites, beginning with the plan for the rebuilding of Plymouth after bombing in 1941, by Patrick Abercrombie. This produced a modernist city centre using a grid street plan. The plan was progressive in its aim to provide a well-designed space for all social classes, but relied on professional expertise and top-down methods. This approach characterises much in modern urbanism, and shaped debates in the International Congress of Modern Architecture (CIAM), which met in Hoddesdon, England, in 1951. Caught between inter-war modernist planning and post-war humanism, CIAM never resolved the dichotomy.

Keywords Plymouth • Modernist planning • Social ordering • CIAM • Lefebvre

In Chap. 3 I discussed Cartesian space; citing Henri Lefebvre's spatial theory, I concluded that the space of plans conventionally used by modern architects and planners is reductive. In this chapter I extend the argument that, despite its idealism, modernist architecture and planning is undermined by this reductivism, and by a reliance on top-down methods. I begin with the *Plan for Plymouth* devised by Patrick Abercrombie in the 1940s. From this I move to tensions within the *Congres Internationaux d'Architecture Moderne* (International Congress of Modern Architecture, or CIAM), meeting in England in 1951. Finally, I return to Lefebvre's critique of Cartesian space to reconsider progressive and regressive currents in modernist urbanism.

After the Bombing

Plymouth, a naval port in Southwest England, was heavily bombed in March 1941. Fifteen hundred houses were destroyed, and most of the city centre turned to rubble. A thousand civilians were killed, and three thousand injured. Plymouth was the most devastated city in England, providing a blank ground for what became a complete modernist city centre. As architectural critic Owen Hatherley remarks, 'in the British city more damaged than any other by Luftwaffe attacks,' rebuilding produced, 'one great ensemble.'[1]

To raise people's spirits after the bombing, or at least redirect attention from bomb sites, a band played on the Hoe (the grassy space looking out to sea where Francis Drake played bowls before fighting the Spanish Armada in 1588). The King and Queen made a visit. John Reith, Minister of Works in the wartime national government, watched dancing in the open air,

Two thousand people were dancing ... Below them was spread the awful havoc lately wrought on their city; not far away across the sea the enemy. As they danced the summer evening into night I saw a coastal forces flotilla steam out from their Devonport anchorage in single line ahead; there was business for them to do, and they would probably do it all the better for what they could see on the Hoe.[2]

Reith knew that Navy families had lived in the bombed houses and that forces morale would be better maintained if life was seen to go on normally. But it was a show: people bombed out of their houses huddled in cellars at night; and the dancers were from outlying villages.[3] A few days later the Navy began to clear the rubble, and the city council opted to demolish most remaining structures, declaring them unsafe to create a vacant 114 hectares. This vast site provided an opportunity to build a city centre symbolising the better future for which the war was being fought.

Patrick Abercrombie was appointed to draw up a plan, with the City Engineer James Paton Watson and the City Architect Edgar Catchpole. Abercrombie was England's leading planner, already working on a plan for post-war London. The local political elite, Waldorf Astor, the Lord Mayor, and his wife Nancy Astor, the city's Member of Parliament, saw the prospect of a radical new plan for the city as reflecting progressive Conservatism, and made viable by the appointment of such a leading figure to oversee it.[4]

Prior to the war, Abercrombie was a planning academic at London University. He advocated a unified approach between different government departments, and national guidance to limit divergences of intention between local authorities. During the war, he began work on the *County of London Plan*, aiming to address traffic congestion, bad housing, lack of open space, and a legacy of overlapping zoning. Gerald Dix, an assistant to Abercrombie, summarises retrospectively,

> The principal defect over housing was seen to be its general drabness rather than the extremely high density found in some other cities; there was a lack of industry to parallel the social individualism of the separate communities … [which] could be overcome by the social and functional identification of the various centres, each based on a former village or market town … Abercrombie thought that this concentration on community was the most distinctive single contribution of the County Plan.[5]

The same applied to Abercrombie's approach to Plymouth, which offered an unparalleled opportunity to start from scratch.

Collaborating with Watson and Catchpole, Abercrombie introduced a grid plan centred on an axis from the main railway station to the Hoe. This is the 61-metre wide, pedestrianised Armada Way, with landscaping in its centre, from which cross-streets were to allow vehicle access to shops (Fig. 4.1). Buildings along Armada Way and the cross-streets are mainly four or five storeys but a later addition, the Civic Offices block (now sold for redevelopment) has thirteen floors over a two-story base. Department stores and an insurance company took on key buildings. A bank designed by B. C. Sherren provided another visual node (Fig. 4.2).

Hatherley notes that the bank's, 'striped classical columns and Scandinavian blue-tiled clock tower are remarkably similar to the precisely contemporary Finland Station in Leningrad.'[6] The balance of opulence and sobriety here denotes a modernism tuned to social democratic politics: designed for a mass public yet retaining the aesthetic quality hitherto restricted to elite buildings. Another example is the Royal Festival Hall, designed by the London County Council's Chief Architect Robert Matthew with Leslie Martin and Edwin Williams, opened during the

Fig. 4.1 Plymouth, Armada Way. (Author's photograph)

Festival of Britain in 1951. The *Plan* was published in 1943, lavishly printed for wartime, with fold-out colour maps. Armada Way was to be a vista for public enjoyment.[7]

In 1945, Michael Foot, orator, literary historian and future Labour Party leader, was elected MP for Plymouth Devonport. On reading the *Plan for Plymouth*, he said, 'We really can have the most beautiful city in the world.'[8] Foot's agenda was more social than architectural, but architectural historian Alan Powers writes that social improvement was an 'unofficial war aim,' while the reconstruction of Plymouth not only raised morale but also realised the aims of, 'left-wing campaigners.'[9] I am unsure whom he means, but the *Plan for Plymouth* was contextualised by the Welfare State set up by the Labour government in the late 1940s.

Blogger and housing campaigner John Boughton writes of the *Plan* as addressing a Britain, 'whose antique streetscapes and cramped lives were seen by many as tired and obsolete,' which required not only rebuilding but also reimagining; and of the, 'grand elevations … [of]

Fig. 4.2 Plymouth, bank building. (Author's photograph)

Portland stone-faced temples of retail, finance and commerce,' among which, in the 1950s, the largest premises was that of the Coop, with a café, 'where dance bands played.'[10] I see this as a significant step beyond nineteenth-century liberal reformism, which improved the material conditions of the working class in part from humane motives but equally as means to increase productivity and prevent social unrest. Those who benefitted were expected to know their place and be grateful. In the 1940s, after the realignment of the class structure in the inter-war years, and women's part in the war effort, a more egalitarian view prevailed. But Dix notes Abercrombie's, 'love of the *beaux arts* solution for the grand occasion.'[11] And Broughton remarks, 'Abercrombie's inspiration owed most to the Beaux-arts imperialism that shaped New Delhi and Canberra.'[12] The inscription of the Plan on a vacant site, and its sweeping geometry does resemble the planning of colonial cities, and Powers notes the top-down power relations involved in a site, 'viewed as a tabula rasa, even if the legal issues of ownership and compensation were still inscribed on the slate and took much legal effort to resolve.'[13] The *Plan*, further, instituted a clear pattern of zoning, differentiating a centre of shops, offices and administrative buildings from an outer zone of housing. The peripheral housing estates are, like the city centre, of high specification, with generous plots, mainly semi-detached 3- or 4-bedroom houses with front and back gardens, reducing pre-war housing density. But while the aim was to build a better world, the means to so included undermining factors.

This reveals a wider current of instrumentalism which I see as contradicting the beneficial, or Utopian, aims of international modernism. I summarise these factors as,

- From inter-war modernism, the politically neutral but strategic role of professional planners, removed by virtue of their expertise from ideological tensions, working across any political spectrum as if outside history and above everyday life;
- A functionalist ideology in single-use zoning, and a separation of planning and design from construction and occupation, continuing in the post-war period;
- Related differentials of social status, access to information, and agency between professionals and dwellers.

Looking Forward

In the 1940s, the focus was on building a better world after the experience of catastrophe. This added impetus to the quest for the new, in its own right, beside practical issues such as the dereliction of the pre-war housing stock: cramped and damp, lacking unshared toilets or bathrooms. The war had been a chasm between old and new worlds, this time not simply as a historical leap to a world without war (as the 1914–18 war was meant to be) but a leap to a society of mass ease produced through mass production and modern technologies.

This perspective appears in a guidebook for an architectural exhibition in Poplar, in London's East End, as part of the Festival of Britain. The exhibition centred on the Lansbury Estate as an example of new housing in two-story terraces and four-story blocks. The guidebook asks rhetorically whether people would like to see a slag heap at the bottom of their garden and children playing in dirt. While past blight resulted from, 'the mistakes of one hundred and fifty years of a free-for-all philosophy and a policy of I'm alright Jack!'[14] (echoing a literary depiction of industry as smoke and dirt), the future was to be found in planning.

Matthew Arnold had argued against the same I'm alright Jack culture in *Culture and Anarchy* (1867). He saw culture as the only defence against social and moral disintegration (anarchy) in a period of fractured religious allegiances, and as the preserve of the educated middle class. In 1951, planning ushered in a democratised but not dissimilar vision, although the Welfare State also provided for universal access to education and cultural experiences.

Returning to the guidebook, the industrial revolution provided better communications and sanitation, and the benefits of mass production, but, it comments, at a cost of, 'dirty, drab, congested towns, squalid water-fronts, polluted rivers, scarred and blackened countryside ... [while] All this is traceable to lack of planning.'[15] Planning, it says, is thinking ahead, not to produce dull uniformity but to release individuals from squalor:

> Much has been done to put mattes right. Much more still remains to be done. It is slow and costly, but it pays dividends in health and happiness ... The Festival of Britain gives us a breathing space ... a tidying-up campaign has been carried out on a national scale ... This national spring-clean is part of the function of architecture, the real purpose of which is to achieve elegance and order, efficiency and gaiety in our everyday surroundings.[16]

This 1951 guidebook may seem a detour but I think it sums up the aspirations of rebuilding in bombed cities such as Plymouth, Coventry, Hull, Sheffield, Swansea and many others.

In Plymouth, the blank ground of demolition enabled a new vision of what a city—the city—should be; light surfaces and broad spaces, easy movement and fresh air, a high aesthetic combined with practical needs. Abercrombie speaks in Jill Craigie's film *The Way We Live*: 'Plymouth needs pale colours to respond to the sunlight ... What is needed is a city to cheer people up.'[17] He adds that it needs an interesting skyline of flat and vertical masses. These were faced with a mix of Portland stone and cement, with much sculptural detailing. It was an amazing achievement, but did not meet universal acclaim. Architectural critic Charles Hussey called it, 'autocratic regimentation,'[18] and *The Architects' Journal* saw the scheme as, 'a monument to the town planning ideals of the thirties and forties.'[19] That hints at the link between international modernism and Socialism. More broadly the juxtaposition of socially inclusive aims and an expert-led method separating planning and design from building and use (as said above) allows the notion that *a better world can be engineered.*

But the modernist city centre produced by the 1941 *Plan for Plymouth* remains a monument to the post-war welfare settlement. Hatherley writes of a growing loss today in British cities rebuilt after 1945: 'the socialist spirit that impelled us to redesign our chaotic, profiteering cities as something unified, public and civic, without gates or fences or hierarchies. The centre of Plymouth is one of the UK's most spectacular places to feel this.'[20] He sees Armada Way as a space, 'which subsequent planners have tried to efface by dint of everything from funfairs to ... giant TV screens.'[21] In Plymouth, part of the landscaping was removed to make a bare plaza with pencil-like trees at its edges in the 1990s, and the city centre put under a separate public-private management organisation. The giant TV screen is sited here. But I am more sympathetic to the screen than Hatherley, seeing the space well used for major sporting events, with temporary seating and tea stalls. The fairground rides and Christmas markets, too, are welcomed by residents, although they encroach on the city centre's pure modernist design. So, while I could easily (a retired academic in an architecture school) look aghast at the screen and the funfair rides, I see them as another layer, reasserting emotional ownership (if through the limited scope offered by market economics) of the site which is also, but in another way, a vision of an optimistic, socially democratic society, providing the best quality of built environment for all citizens. As Hatherley writes,

> Planned post-war Plymouth is now being recognised as having value ... It's about time that social democratic Britain was the subject of something more than giggling and ridicule, and there's no doubt that the incremental demolitions of decent buildings around the edges of the place and their replacement with dross should be stopped.[22]

The dross is the piecemeal, opportunistic redevelopment of sites with steel-and-cladding hulks, often for student housing in very small rooms, in a fantasy of regeneration.

DESIGN FOR A HUMANE WORLD

In the 1950s, signs of economic recovery began to appear and bomb sites were cleared for rebuilding in British cities. In Germany the bombing had been worse—the annihilation of civilian populations as well as old city centres in Hamburg and Dresden, for instances—and reconstruction tended to a modernist pattern as a renewal of values which the Nazi regime, with its regressive culture of blood and soil, lost crowns in the Rhine and songs around the forest campfire, had trashed.[23] Remembrances of social harmony competed with forward-looking daring in urban planning and architectural design, informed by a humanism which was less a secular religion than a concern for the basic bonds of humanity. This was CIAM's terrain. Design historian Barry Curtis writes of CIAM,

> [This humanism] was deeply imbued with notions of continuity and tradition, and marked by the terms Man or Modern Man. Younger artists and designers of the 1950s sought alternatives to this formulation in their admiration of a culture of consumption which recognised the fragmentation of the market and the indifference of the dominant humanist way of thinking. Humanism was an organising principle in architectural thinking in the post-war period ... but [it] also implied a new mode of sensitivity to values which responded to recent experiences of totalitarianism and scientifically planned mass destruction.[24]

Another humanism appears in post-war psychiatry when John Bowlby argues for attention to both social reconstruction and individual desires and routines. Among such desires were home ownership, watching sport, and seaside holidays. He asks how to reconcile personal desires with,

> the understanding and acceptance of the need for the inevitable controls required for the attainment of group goals such as, for instance, full employment, a maximising of production ... or a maximising of personal efficiency through longer and more arduous education and other social measures.[25]

In Britain, this was against a background of rationing until 1954, and mass re-housing. The Welfare State introduced the National Health Service, and extended pre-war social security; it also relied on government planning, and a uniformity of services across the country (much as Abercrombie sought through the introduction of national planning guidelines).

Sociologist David Kynaston cites Michael Young, author of the 1945 Labour Party manifesto, that the need was for both top-down planning and grass-roots democracy, to undertake the 'ten-year plan' on which the new government had stood at election; but since the plan was not widely understood, 'it was dangerous to wait and hope for the best … members of the Party must themselves get going and assist the process.'[26] The process required vision and means, or design and the technology, money and systems to deliver it. For Young, planning was the prerogative of a qualified elite and acceptance of the plan was delegated to a mass public. For people occupying new homes with indoor toilets and hot water this was not a problem. Journalist John Grindrod reports a wealth of social research affirming occupants' delight at moving into new houses, or flats with plate glass windows and a view. On Park Hill, Sheffield, for example, he says, 'almost everyone seemed happy with this spectacular and intimate new Park Hill.'[27]

In London, some people were relocated to prefabricated one-storey houses, built under the 1944 Housing (Temporary Accommodation) Act, called prefabs. Experimental prefabs were exhibited at the Tate Gallery in 1944, one designed in-house at the Ministry of Works using a steel body to be produced by car manufacturers; another came from a manufacturer in Hull, using re-enforced concrete panels; a third was from a mass housing builder, with a pitched roof; and a fourth was from an engineering company, timber-framed with asbestos cladding. Thirty thousand of the latter were produced, some of them still occupied in 2012 when Grindrod met residents of the Excalibur Estate in south London. He quotes one who moved into a prefab after being demobbed from the forces. With a wife and two children, in bomb-damaged property, at first he refused it, until,

> They told me to go and look before I decided. We opened the door and my wife said, "what a lovely big hall! We can get the pram in here." There was a toilet and a bathroom. I'd been used to a toilet in the garden. The kitchen had an Electrolux refrigerator, a New World gas stove, plenty of cupboards. There was a nice garden. It was like coming into a fortune.[28]

This reflects the post-war housing shortage but the names—Electrolux, New World—convey optimism. Grindrod notes that many of the prefabs on the Excalibur Estate were repainted in pastel colours; but also that prefabs were built by German and Italian prisoners of war working as 'forced labour.'[29] When he visited the estate in 2012 he saw notices in windows saying, 'I'm not moving,'[30] against efforts to replace these sixty-year-old temporary homes.

Prefabs did not need architects but more than two hundred architectural practices entered a competition for the design of a new Coventry Cathedral (the old destroyed in bombing) in 1947. Among them were Basil Spence (who won) and Alison and Peter Smithson, known for their Brutalist designs in the 1960s. Grindrod describes their proposal as, 'something that was both as high tech as jet planes and as starkly monolithic as Stonehenge: a vast kite-like shelter that soared above a platform containing the open box of the chapel.'[31]

In the new Welfare State of post-war Britain, this combination of future vision and ancient monument memorialised an indefinite past lit by a light shining from the end of history onto a fractured present. Architectural historian Joe Kerr writes,

> In a very real sense, the houses, hospitals, and schools that came to dominate the landscape of London were ... the expeditionary force for a great new campaign. Aneurin Bevan, opening a new London housing estate in 1948 spoke of such buildings as if they were anticipatory fragments of an equitable society of the future ... [so that] the city and its architecture became the peaceful monuments for peacetime battles.[32]

The idea of peacetime struggles placated by peaceful edifices informed CIAM's debates in the 1950s. The edifices were public and domestic, symbolic interventions in an urban fabric, and settings for ordinary lives. The difficulty, again, is that the combination of the symbolic and the ordinary reproduces the conflict between engineering a new society and its social production.

CIAM's concerns included modern civic ordering, but also the kinds of civic relations which instantiated that order. As Curtis notes, architects were expected to envisage a future which people could accept. CIAM's members were wary of central planning, although some had worked on projects in the Soviet Union (an ally in the defeat of Fascism). Curtis continues that a recurring injunction was to, 'respect and remedy the causes of totalitarianism and authoritarian thinking' which might be done either by inclusive planning or, 'by abandoning the concept of planning altogether.'[33] Architects were visionaries tasked with foreseeing the better world imagined in the post-war climate of opinion but could they also be agents of democracy without resigning their privileged status, and levelling the claims of public and domestic buildings?

Boughton cites the *Plan for Plymouth* as dealing with, 'the comfort and convenience of the smallest house,' and, 'the magnificence of its civic centre, the spaciousness and convenience of its shopping area and the perfection of the industrial machine.'[34] Abercrombie saw social renewal as produced by a new built environment: better spaces produce better citizens who contribute to a better world. But, Boughton comments,

> The utopian hopes betray the temper of the times as least as much as the gendered language. Planners, who believed themselves to be unleashing such human potential and who themselves ... personified sweet reason and benevolent intention, can be forgiven for thinking themselves central to the construction of this new world.[35]

By the 1950s, the modernist styles of the inter-war period had become normalised in civic buildings while the extent of rebuilding allowed an application of modern design in housing estates. The brave new society was being manufactured in both inner-city reconstruction and peripheral estates, and in new towns such as Milton Keynes, Harlow and Peterlee. The guidebook for the 1951 architectural exhibition (cited above) shares the optimism but is not unaware of diverse interests, which it sees as a difficulty:

> Building new towns makes a fresh start possible. ... We can all see the need for properly designed road systems with no dangerous crossings or corners, for plenty of open space for recreation, and of course for up-to-date housing sited away from industrial areas and grouped within easy reach of schools and shops.
>
> The difficulties ... begin when the many and sometimes conflicting needs of people of different callings, ages and backgrounds have to be provided for in the plan.[36]

In West Ham, Abercrombie's plan for London was implemented by local authority architects in consultation with local politicians. Architect Nicolas Bullock notes, 'This was to be no utopian blueprint foisted onto an unwitting public by an arrogant architect/planner, but a set of proposals developed ... in conjunction with West Ham's leaders.'[37] It nodded towards democratic planning but in a representational politics subsuming individual voices in public expressions of value which might or not be shared.

DIVIDED SPACES

CIAM was founded in 1928 in a Swiss chateau. Among the twenty-eight modernist planners and architects who participated were Le Corbusier, Sigfried Giedeon, Karl Moser, Ernst May, Hendrik Berlage, Gerrit Rietveld and Hannes Meyer. A Soviet group included Constructivist artist El Lissitzky. Members joining later included Walter Gropius, Josep Sert and Alvar Aalto. CIAM's participants represented the spectrum of modernist architecture from the Bauhaus in Weimar Germany to De Stijl in the Netherlands and Sert's modernism in Barcelona. There was also a North American branch, the Chapter for Relief of Post-war Planning, in New York. CIAM was concerned with city planning, but also with housing. Its third meeting, in 1930 in Brussels, 'Rational Land Development,' debated the competing merits of high- and low-rise development. Gropius and Le Corbusier argued for high-rise, May for low-rise. But all were united in advocating functionalism: form follows function; and space is aligned to specific functions.

Le Corbusier combined single-use zoning with a separation of motor and pedestrian traffic, on a similarly delineated scale from public to domestic life. Everything is in its place. Using a universal masculine, he asserts in *The City of Tomorrow and its Planning,*

> The house, the street, the town, are points to which human energy is directed; they should be ordered, otherwise they counteract the fundamental principles round which we revolve; if they are not ordered, they oppose themselves to us ...
>
> If I appear to be forcing an already open door ... it is because ... certain highly placed persons who occupy strategic points on the battlefield of ideas and progress have shut these very doors, inspired by a spirit of reaction and a misplaced sentimentalism which is both dangerous and criminal ...
>
> I repeat that man, by reason of his very nature, practices order; that his actions and his thoughts are dictated by the straight line and the right angle; that the straight line is instinctive in him and that his mind apprehends it as a lofty objective.
>
> Man, created by the universe, is the sum of that universe ... he proceeds according to its laws and believes he can read them; he has formulated them and made of them a coherent scheme ... on which he can act, adapt and produce.[38]

This is essentialist and anthropomorphic—humans self-evidently exemplify the order of the universe—and combative—a battlefield of ideas—while seeing opposition (to his ideas) as criminal. It is also gendered and prescriptive. Little wonder that Le Corbusier's *Voisin Plan* for Paris (named after a sponsor) was first published in the French Fascist newspaper. This emulation of a Cartesian reliance on closed systems departs, however, importantly, from Descartes' intention to find a method only for his own enquiry (Chap.3) by universalising design as inscription.

CIAM's fourth meeting, in 1933, was planned for Moscow; but when Le Corbusier's design for the Palace of the Soviets was rejected—it was won by Boris Iofan—the meeting was relocated to a ship, SS Patris, sailing from Marseilles to Athens. This meeting debated 'The Functional City,' concluding that cities should be zoned, their populations housed in widely spaced tower blocks. This is more or less what Le Corbusier proposed in 1929, in a formula for a city of three million people, on a level site, with a river as a, 'liquid railway,' or industrial station: 'in a decent house the servants' stairs do not go through the drawing room – even if the maid is charming.'[39] Leaving aside his intentions toward the maid,[40] the remark shows a hierarchical outlook in excluding servants from the sight of the property's owners, which is equivalent to single-use zoning's exclusion of peripheral domestic from central public space. Both are separations based on an abstract system; and both, despite the planner's neutral guise, are expressions of privilege. Again, I read this as indicative of the undermining undercurrent in modernist urbanism, the autocratic voice expressing democratic ideas, the instrumentalism which denies a professed belief in society.

In *The City of Tomorrow*, spatial and status distinctions are combined. Le Corbusier lists, in order of importance, citizens 'who work and live in [the city],' suburban dwellers who work in an outer industrial zone and, 'do not come into the city,' and a 'mixed sort' who work in the city but live with families in satellite garden cities.[41] The street should be a, 'masterpiece of civil engineering,' while mixed-use, 'corridor,' streets, 'should be tolerated no longer.'[42] City dwellers live in towers at a density of 1200 people to an acre (although inhabitants of luxury dwellings enjoy a tenth of that); the city centre has cafés, restaurants, luxury shops, and a forum adjoining an 'immense' park, 'providing a spectacle of order and vitality.'[43] It sounds like the Garden City (Chap. 2), or a mall. Looking at his profession in a jaded light, Le Corbusier writes, 'The architect ... has become a twisted sort of creature. He has grown to love irregular sites ... Of course, he is wrong.'[44] Le Corbusier saw himself as primarily a planner, and was a, 'looming presence', at CIAM 8 in England in 1951.[45]

THE HEART OF THE CITY

Hoddesdon was an odd place for CIAM: a small market town, once a coaching stop between London and Cambridge, distinct only in having a large community of Italian ex-prisoners of war. CIAM met there in 1951, just before the fall of the Labour government. Its proceedings were still haunted by Fascism and the devastation caused by wartime bombing, seeing city planning as defence against the recurrence of catastrophe. Its title, 'The Heart of the City,' suggests a democratic core value and, as Curtis puts it, 'the construction of contexts which would stimulate citizens to activity.'[46] This included designing civic centres but was also an intellectual agenda differentiating mass society, the crowd, from motivated individuals; or not, in the view of some younger members. Curtis draws an analogy with cinema:

> The spatial criteria of the conference resemble the priorities ... for quality cinema in the immediate post-war period. The new cinematic realism ... [used] shots from the point of vantage ... individualised and representative images of ordinary people and a narrative and deeply focused *mis en scène*.'[47]

Among the ordinary people were, in Ian McCallum's presentation, flower girls and soapbox orators, and the hand-written blackboards advertising the day's catch in fishmongers' shops. This was the informal, everyday city, beyond design but with its own culture. For some other presenters, architecture's expert status was non-negotiable. A universal masculine seemed unquestioned. Curtis notes, 'Women played little part ... Jacqueline Tyrwhitt, the secretary, seems to have been the only female member of the council.'[48] Sexism is compartmentalising (as is single-use zoning) while zoning constitutes a total-Cartesianism. A difficulty emerges in the privileged status lent by CIAM 8 to the classical Greek *agora*, too, as an a-historical, de-politicised ideal supposed to stand for the ideal city.

CIAM's *agora* is a post-war projection, reacting to totalitarianism by projecting an image of modern democracy onto a remote historical site. In fact the *agora* of classical Athens was a market surrounded by booths for administrative uses, and a colonnade where free citizens met. Some public events occurred there but most political decisions were taken, by rich city-born men, in the Assembly (*pnyx*). Regardless, for CIAM, the *agora* had a mythic status. I think this romantic view occurs, too, coincidentally, in

Jane Jacobs' picture of streets as sites of informal mixing (Chap. 1). For CIAM in 1951, it supported post-war humanism, as Curtis explains,

> Humanism in architecture is closely associated with anthropomorphy and empathy. It assumes a unified subject and a fundamental need for compositional integrity. As it engaged with the humanistic culture of post-war Europe, it focused on a future which would see fulfilment of human potential, a future predicated on a past of essential humanity which had to be retrieved from the superficial accumulations of everyday life.[49]

Meanwhile, Existentialism in Paris insisted on the immediacy of human action and decision, an agency of engagement which could not, after the 1930s, rest. And Lefebvre was writing of cities as sites of everyday life and momentary but transformative insights.

For CIAM, humanism stood for an organic approach—against which some younger members reacted in a call for sociological research—which was paradoxically a departure from rigid spatial programming *and* an affirmation of the need for a unifying method. For Curtis it was a juncture between Giedion's idea of the organic plus, 'the rational and geometric,' and the position proposed by another participant, Bruno Zevi, of humanism as disallowing aesthetic dogmatism. Curtis sums up that organic architecture could, 'describe both processes of ordering and of *laissez-faire*.'[50] The question is on what basis intervention is to be made.

I gave the last word on Plymouth to Hatherley, and give the last word on CIAM 8 to Curtis, writing in 2000 after the rise of new cultural and social theories since the 1960s:

> Perhaps the most constructive way of rethinking the troubled discourse of CIAM 8 is as part of a dialectic between the claims of a return to stable or absolute verities and an opposing desire to recognise a field of competing values. ... It is clear that the dominance of structuralist and poststructuralist theory has ... repressed value at a time when it has become a key issue in everyday politics. Clearly there is no way back to the essence and indifference of the early 1950s ...[51]

LIVED SPACES?

Poststructuralism took apart the orderly explanations advanced in modern sociology and anthropology, and in the humanities, to show that, far from revealing how reality is, over-arching explanations (meta-narratives) reflect an author's preconceptions. For urbanism, this means that the eye-in-the-sky of the conventional city plan disposing regular spaces on a blank ground represents a god-like viewpoint—the visual sense as mastery, as geographer Doreen Massey put it[52]—and not the complex, plural realities of city lives. The lesson is that there is no objective perception of reality, only as many subjective perceptions as there are viewers (like brush-marks in a painting by Paul Cezanne). Impressions can be summarised for convenience but the conditions summarised and the resulting summaries are contingent in both cases on the conditions in which they occur, as are words in a system of language.

A plan for urban renewal is thus a historically specific product of economic, political, cultural, and social conditions, and of power relations within those conditions. The universalism of Le Corbusier's segregation of classes of urban dweller, then, which he presents as universal, as if it can be carried out anywhere and is self-evident, is untenable. But, if it is easy to write off Le Corbusier's modernism as totalitarian, it can also be read as a polarity on an axis from ordering inscription to co-production, implying a large grey area between the two, where the work of urbanism is done. This is easy to write but the processes of living and working, of negotiating urban spaces, is not simple, drawn between instrumentalism and agency in a terrain of permanent contest and contestability.

An instrumentalist approach sees the ends as justifying the means (under certain regimes this becomes any means); with agency, citizens claim and contest space, voice and visibility, arriving at ephemeral resolutions. Modernism draws a Utopian image of a city but makes it *the* city; yet an imagined better world can also be seen as a work in progress, its form following enactment of its values rather than a set of functions, mirroring the difference between the representational democracy of modern nation-states and direct democracy. Good design is, in this respect, a process of evolution (just as verbal language adapts to received regulation and everyday usage, in both directions at once).

I remember being at a conference on urbanism at Tate Liverpool in the 1990s at which there was a lively discussion on Utopianism. Towards the end someone said, to paraphrase from memory, 'but someone has to design it,' met by a resolute 'no' from a person on the Left.[53] I am reminded, too, of the conflict between housing design and adaptation at Pessac where Le Corbusier was asked to design a garden city in 1926. The houses were to be of unusually high specification for the time, with central heating, hot water, wide windows, electricity, showers, and a mix of balconies and small gardens. There were differences between Le Corbusier and the industrialist Henri Frugès who commissioned the development. Frugès recalls,

> With his [Le Corbisier's] inveterate hatred of all forms of decoration (which stemmed from his Protestant background and the general austerity of his personality) he wanted to leave the walls completely unfinished so that they still showed the marks of the shuttering [the wood surfaces used to form concrete sections]. I was flabbergasted. It was in vain that I asked him to put himself in the place of the future purchasers, whose eyes are accustomed to decorative effects ... At that moment ... [I saw] the idea of painting the façades of the villas in different colours ... so that they would harmonise with one another and also be visible ... from the other side of the green areas.[54]

Le Corbusier was also a painter, and accepted the idea. The outcome was a development of low-cost housing in two-story villas and four-story apartment blocks in bright colours, with balconies and in some cases exposed outer stairways. But residents adapted the properties, filling in open sections, adding porches, workshops or car ports, making interior alterations as well. These were DIY interventions. Social researcher Philippe Boudon summarises that, in the late 1960s, the occupants voiced functional, rational, and aesthetic reasons for having changed the design but, 'were often motivated by quite different considerations, which were based on established standards of taste.'[55] In other words, dwellers familiarised their houses at the cost of pure design. The buildings have since been restored for architectural preservation (or an architectural museum with living inmates, perhaps).

This zone of claim and counter-claim sits uneasily beside the supposed universality of spatial conditions in functionalism, which latter Lefebvre observed in the new town of Mourenx near his semi-rural home:

> Mourenx has taught me many things. Here, objects wear their social credentials: their function, every object has its use, and declares it. Every objet has a distinct and specific function. In the best diagnosis ... everything in [Mourenx] will be functional, and every object in it will have a specific function, its own. Every object indicates what this function is, signifying it ... It repeats itself endlessly. ... a group of these objects becomes a signalling system. ... The text of the town is totally legible, as impoverished as it is clear.[56]

Later he says the world is made to be like a Meccano set (metal parts for assembly into any structure, a children's toy popular in the 1950s).[57] Near the end of this essay he juxtaposes, 'spontaneous vitality, and abstraction,' the former aligned to pleasure and play, the latter to toil and routine.[58] Characteristically, Lefebvre does not pose this as a dualism,

> And here we are facing the same problem as before: how to reproduce what was once created spontaneously, how to create it from the abstract. Possible? Impossible? If the concept of what is possible is separated from the concept of what is not possible, both become meaningless. If we aim at what is impossible today, it will become tomorrow's possibility. And no matter how immense the gap … the aim and the goal, abstraction and living, may appear, you must make every effort to bridge it.[59]

Or, from his later theory of space, conceived space and lived space (representational space and representations of space) are polarities on a dialectical axis within a society's spatial practices. Each is always there, modified by the other.

I return to Lefebvre's theory of planned space and lived spaces now because modification is a characteristic of a lived space. Le Corbusier produced planned space, ideally inscribed on a blank site. Lefebvre takes sites of occupation as having lives of their own, as it were, so that lived space, while dominated by planned space, tends to reappear and interrupt the ideal but brittle uniformity which can never quite be maintained. From this, follows a dialectic of design and use on a potentially creative axis. But how does it work?

I recall an interview for a job I did not get at which I cited the ability of dwellers to embellish environments as a mark of emotional ownership. One of the panel retorted, 'Do you mean stone cladding?'[60] I didn't, but could have because stone cladding—a veneer of stone effect applied on brick or concrete walls—is one of few available means open to, say, people who have recently bought their council house and want to personalise its exterior appearance. I am not defending the Right to Buy scheme which enabled that in Thatcher's Britain; and did not say I liked stone cladding. But embellishment can be understood socially, while for the panel member (from the Philosophy Department) it was unacceptable, mere bad taste.

Emotional ownership produces adaptation, an ephemeral re-functioning of space. In public spaces it can be using a step as a seat, or moving a portable chair in a café. In a perspective from above, this is dirt—matter out of place—yet from a street-level position it is everyday life. Thinking of Pessac, I am reminded that architectural historian Simon Sadler writes,

> Modernism had once assumed that revelation would be achieved by contemplation of the ... ideal architectural object, but non-planning promoted the notion of architecture as event and situation ... realised by the active involvement of the subject.[61]

By non-plan, Sadler means a non-formal, creative approach to design as a continuing work of making and re-making, and re-making again, spanning from experiments by groups such as Archigram in the 1960s to, 'the guerrilla architectures of riot, squatting and nomadism.'[62] I might put it, instead, as the occupational architecture of anti-roads or anti-airport protests in the 1990s, from tree houses to tunnels, or the informal shed-settlements put up by some Occupy groups in 2011–12; but, equally, permanent self-build housing which negotiates a role in planning regimes but not in market economics.

Perhaps the idea which lingers is architecture as event; and from that space as produced, not given. Lefebvre writes of social space that it is a palimpsest of occupations and traces, understood through aspects of ethnology, ethnography, human geography, anthropology, prehistory and history, and sociology. Then he says,

> In the immediacy of the link between groups, between members of groups, and between society and nature, occupied space gives direct expression ... to the relationships upon which social organisation is founded. Abstraction has very little place in these relationships.[63]

Those relationships are not power relations (or relations of power-over) but empowerment (power-to). But this is not to say that abstraction, in the form of imagining a possible future (even if thought impossible at the time or by critics), is absent; only that the dream is one of the factors, and various practicalities, including political struggle, are others. The question is to what extent the dream becomes a prescription, or evolves as an invitation.

NOTES

1. Hatherley, O. (2012) *A New Kind of Bleak: Journeys Through Urban Britain*, London, Verso, p. 179.
2. Reith, W. (1949) *Into the Wind*, London, Hodder and Stoughten, p. 428, quoted in Hall, P. (1996) *Cities of Tomorrow*, updated ed., Oxford, Blackwell, pp. 219–220.
3. Wintle, F. (1981) *The Plymouth Blitz*, Bodmin, Bossiney Books, p. 56, quoted in Powers, A. (2002) 'Plymouth: Reconstruction After world War II,' Ockman, J., ed., *Out of Ground Zero*, Munich, Prestel Verlag, p. 100.
4. Gould, J. (2010) *Plymouth: Vision of a Modern City*, Swindon, English Heritage, p. 5.
5. Dix, G. (1981) 'Patrick Abercrombie,' Cherry, G.E. *Pioneers in British Planning*, London, Architectural Press, p. 115.
6. Hatherley, *A New Kind of Bleak*, p. 181.
7. Gould, *Plymouth*, p. 7, citing Abercrombie, P. and Paton Watson, J. (1943) *A Plan for Plymouth*, Plymouth, Underhill, p. 67.
8. Foot, M. (c.1945) from Hoggart, S. and Leigh, D. (1981) *Michael Foot: A Portrait*, London, Hodder and Stoughton, p. 91, quoted in Powers, 'Plymouth,' p. 108.
9. Powers, 'Plymouth,' p. 101.
10. Boughton, J. (2019) *Municipal Dreams: The Rise and Fall of Council Housing*, London, Verso, pp. 62–63.
11. Dix, 'Patrick Abercrombie,' pp. 120–121.
12. Boughton, *Municipal Dreams*, p. 63.
13. Powers, 'Plymouth,' p. 102.
14. Dunnett, H.M. ed. (1951) *Guide to the Exhibition of Architecture, Town-Planning and Building Research*, London, H.M. Stationery Office, p. 5.
15. Dunnett, *Guide*, p. 5.
16. Dunnett, *Guide*, p. 5.
17. Cited in Powers, 'Plymouth,' p. 106.
18. Hussey, C. (1944) 'the New Plymouth,' *Country Life*, 12 May, pp. 812–813, cited in Powers, 'Plymouth,' p. 109.
19. Editorial, (1952) *Architects' Journal*, 115, June, p. 715 cited in Powers, 'Plymouth,' p. 112.
20. Hatherley, *A New Kind of Bleak*, p. 178.
21. Hatherley, *A New Kind of Bleak*, p. 179.
22. Hatherley, *A New Kind of Bleak*, p. 188.
23. Bloch, E., [1934] (1991) 'Rough Night in Town and Country,' *Heritage of Our Times*, Cambridge, Polity, pp. 48–55.

24. Curtis, B. (2000) 'The Heart of the City,' Hughes, J. and Sadler, S., eds., *Non-Plan: Essays on Freedom, Participation and Change in Modern Architecture and Urbanism,* Oxford, Architectural Press, p. 52.
25. Bowlby, J. [1945] (2007) contribution to Fabian Society meeting 'The Psychological and Sociological Problems of Modern Socialism,' quoted in Kynaston, D., *Austerity Britain 1945–51,* London, Bloomsbury, pp. 127–128.
26. Young, M. [1945] (2007) quoted in Kynaston, *Austerity Britain,* p. 130.
27. Grindrod, J., (2013) *Concretopia: A Journey Around the Rebuilding of Postwar Britain,* Brecon, Old street Publishing, p. 177.
28. O'Mahoney, E. [2012] (2013) *The Guardian,* 28 December 2012, quoted in Grindrod, *Concretopia,* p. 27.
29. Grindrod, *Concretopia,* p. 31.
30. Grindrod, *Concretopia,* p. 34.
31. Grindrod, *Concretopia,* p. 112.
32. Kerr, J. (2001) 'London, War, and the Architecture of remembrance,' Borden, I., Kerr, J., Rendell, J. and Pivaro, A., eds., *The Unknown City: Contesting Architecture and Social Space,* Cambridge (MA), MIT, p. 80.
33. Curtis, 'The Heart of the City,' p. 53.
34. Abercrombie, P. and Paton Watson, J. [1945] *Plan for Plymouth,* Plymouth, Underhill, p.2, cited in Boughton, *Municipal dreams,* p. 62.
35. Boughton, *Municipal Dreams,* p. 66.
36. Dunnett, Guide, pp. 37–38.
37. Bullock, N. (2015) 'West Ham and the Welfare State,' Swenarton, M., Avermaete, T. and van den Heuvel, D., eds., *Architecture and the Welfare State,* London, Routledge, p. 95.
38. Le Corbusier [1929] (1987) *The City of Tomorrow and its Planning,* New York, Dover, pp. 15–17.
39. Le Corbusier, *The City of Tomorrow,* p. 165.
40. See Colomina, B. (1996) *Privacy and Publicity: Modern Architecture and Mass Media,* Cambridge (MA), MIT, pp. 82–100.
41. Le Corbusier, *The City of Tomorrow,* p. 166.
42. Le Corbusier, *The City of Tomorrow,* p. 167.
43. Le Corbusier, *The City of Tomorrow,* p. 172.
44. Le Corbusier, *The City of Tomorrow,* p. 176.
45. Curtis, 'The Heart of the City,' p. 53.
46. Curtis, 'The Heart of the City,' p. 56.
47. Curtis, 'The Heart of the City,' p. 58.
48. Curtis, 'The Heart of the City,' p. 60.
49. Curtis, 'The Heart of the City,' p. 61.
50. Curtis, 'The Heart of the City,' p. 63.
51. Curtis, 'The Heart of the City,' p. 64.

52. Massey, D. (1994) *Space, Place and Gender*, Cambridge, Polity, p. 232.
53. Undocumented personal memory, c.1996–67.
54. Frugès, H. [1967] (1972) quoted in Boudon, P. *Lived-in Architecture: Le Corbusier's Pessac Revisited*, London, Lund Humphries, pp. 9–10.
55. Boudon, *Lived-in Architecture*, p. 81.
56. Lefebvre, H. [1960] (1995) 'Notes on the New Town,' *Introduction to Modernity*, London, Verso, p. 119.
57. Lefebvre, 'Notes on the New Town,' p. 121.
58. Lefebvre, 'Notes on the New Town,' p. 125.
59. Lefebvre, 'Notes on the New Town,' p. 125.
60. Personal memory, c.1993–94.
61. Sadler, S. (2000) 'Open Ends: The Social Visions of Non-Planning,' Hughes and Sadler, *Non-Plan*, p. 151.
62. Sadler, 'Open Ends,' p. 152.
63. Lefebvre, H. [1974] (1991) *The Production of Space*, Oxford, Blackwell, p. 229.

CHAPTER 5

Post-industrial Ruinscapes

Abstract This chapter turns to contemporary Germany and the preservation of industrial sites in the Ruhr. Extensive in scale, these redundant mining and steel-making sites have become monuments, or memorials, to an industrial past which provided technical advances and material benefits in mass production, but was environmentally damaging. The sites are now de-contaminated and re-greened, offering an ambivalent meaning: preserving a dead industrial past while looking to a post-industrial future. In Britain, redundant industrial sites are more often abandoned but are accessed informally by urban explorers for whom they offer a means to reclaim urban space.

Keywords Industrial ruins • Rhur • Urban regeneration • Urban exploring

In Chap. 4 I argued that instrumentalism undermined the Utopian intentions of modernist planning and architecture. In this chapter I move from modern to postmodern culture, and from cities and their planning to the ruinscapes of de-industrialisation in the Ruhr, Germany. I look at these sites because they starkly state the end of the industrial period, and with it an aspect of modernity in mass production and the solidarity of labour organised in large-scale factories. Their re-coding as leisure spaces also puts post-industrial ruinscapes in the context of the symbolic economies

M. Miles, *Paradoxical Urbanism*,
https://doi.org/10.1007/978-981-15-6341-6_5

of immaterial production which now dominate the global North. But I also look at the emergence of urban exploration—non-sanctioned entry into industrial or corporate spaces, mainly disused—as spontaneous reclamations of urban spaces outside the symbolic economy's definition of creativity.

New Wastelands

Post-industrial ruinscapes constitute a distinct landscape: trees, grasses, wild, and cultivated plants surround rusting but preserved industrial structures, in coal mines and coking plants, steel furnaces, and chemical works. The sites are decontaminated for public access but their massive structures remain, evoking a past of skilled labour, the benefits of mass production, or, in a time of climate crisis, pollution. Re-greening the sites lends them a progressive aura, a quality of re-naturalisation after centuries of smoke. But is this a mask? And are the ruins memorials, or pleasure-grounds?

The beginning of the industrial epoch saw new kinds of vista in the landscaped park, offering private contemplation to their aristocratic owners. Post-industrial ruinscapes offer retreat as well, for local people and cultural tourists. But today's democratised leisure is set against economic readjustment, unemployment, and de-skilling[1]; and designed as compensation for routine in a wheel of earning and spending when new desires (or false needs) are produced in order to be only partly satisfied, keeping the wheel of work-and-compensation relentlessly turning.[2] In this context, post-industrial ruinscapes are illusionary Utopias where nature, reframed as culture, as such a seemingly non-contentious domain, compensates for exploitation. Yet they carry a more complex range of meanings, refracting urban modernity via the succession growth of uncultivated flora (brambles and wild flowers which naturally grow on a site).

The Ruhr

The Ruhr in north-west Germany was a region of coal, iron, steel, and chemical production in the nineteenth and twentieth centuries. The end of material production left sites too vast to clear, and which were not needed by new industries or housing in a region with a shrinking population. Reconfiguration of the sites was seen as a path to economic recovery, although that has not been fully or evenly achieved. It was enabled by designating their preservation as an International Architecture Exhibition (*Internationale Bauaustellung*, IBA).

Past IBA projects range from Joseph Olbricht's *Mathildenhohe* in Darmstadt in 1901 to an inner-city project with citizen participation in Kreuzberg, Berlin, in 1979–1987. Today, the redundant towers, turbine halls, gantries, bunkers, and cooling ponds are open to leisure uses, almost as if in a democratised reform of the eighteenth-century landscaped park with its temples, grottoes and diverted water courses. The landscaped park became popular in nineteenth-century Germany, called English Garden, so it seems legitimate to make the comparison. But, as I explain below, it is not an exact comparison, since the temples of a park (such as Rousham, Chap. 2) were built for the purpose of visual diversion, and were aesthetic, not useful. Industrial structures may be lent aesthetic appeal in postmodern eyes but that was never their purpose.

One of the most extensive sites is the combined colliery and coking plant at Zollverein, near Essen. This is a UNESCO World Heritage site, pit 12 having been the largest mine in Europe, designed by Bauhaus architects Martin Kremmer and Fritz Schupp, who incorporated a high level of mechanisation in the sorting and movement of coal. Zollverein was opened in 1932 by Adolf Hitler (who approved of modern design in industrial but not public or domestic buildings). By 1935 the mine supplied 12,000 tons of coal per day to the German steel industry. Kremmer and Schupp are quoted,

> We should acknowledge that gigantic industrial buildings are no longer a blot on the urban landscape but a symbol of work, city monuments, for local citizens to show to outsiders alongside other public buildings, and with as much pride.[3]

Zollverein closed in 1986. The site now houses a design museum, Red Dot (redesigned by Norman Foster), a regional history museum, a restaurant, and an ice skating rink in winter. Birds have returned in the woodlands which have naturally reclaimed the site.

Elsewhere in the Ruhr, the gas blower and dynamo station at Bocchum is a music venue; the nearby Hannover colliery is an industrial museum; Heinrichstütte steelworks at Hattingen is another industrial museum; a disused gasometer at Oberhausen houses an exhibition hall; a sculpture trail connects various slagheaps; art and historical museums occupy the colliery at Dortmund; and a mining site at Gelsenkirchen houses a science centre, woodland park, and monument by artist Herman Prigann (discussed below). A guidebook summarises the Ruhr's current plight:

> Superficially the region seems to be little more than a conglomeration of run-down industrial towns in search of a future, with nothing to offer except beer and football. This was once the area where King Coal reigned supreme ... [Now,] the legacy of filth and pollution left by decades of heavy industry has been largely cleaned up, revealing a surprising amount of beautiful countryside. ... [And] the industrial monuments which were once the centre of people's everyday existence have not only been preserved, but restored in the form of on-site museums, artistic venues and community centres as a living tribute to the men and women whose blood, sweat and tears transformed the region into the powerhouse of Germany.[4]

While the references to beer and football, and blood, sweat, and tears might be aimed at English readers in an English language guidebook, it also points to a history of work and the role of industry in ordinary lives.

Beginning in 1989, the IBA announced seven aims: to reconstruct the landscape; to restore the ecology of the river system; to create an outdoor adventure space; to act as a cultural heritage site; to produce new employment; to introduce new forms of housing; and to offer social, cultural, and sports activities for local people and visitors.[5] Employment was elusive, and efforts to establish a solar power industry were undermined by imports from China at prices which undercut German production. But environmental restoration was successful, leaving de-contaminated (or buried) soil and clean water courses. The IBA justifies the turn to leisure space on the grounds that the sites would have found no other use, and that this offers significant community benefits.

DUISBURG NORD

Duisburg Nord Landscape Park (*Landschaftspark Duisburg-Nord*) occupies the 230-hectare site of the Thyssen ironworks. Its blast furnaces, water tanks, gantries, canals, rail lines, and storage buildings offer a cacophony of steel, concrete, brick, grass, and trees. Duisburg Nord merges into the even larger Emscher Park, along the river Emscher, with waterside paths and further, longer vistas. Spending a day there, I was captivated by the strange familiarity of the fusion of industrial relics and natural growth (with some new planting). The scale of the industrial structures, and their dislocation amid trees, reminded me that Anglo-Saxons saw the remains of roman Britain as the works of extinct giants.

Coal mining began at Duisburg in 1899. The ironworks began production in 1903, generating 37 million tonnes of iron over its lifetime. Stopped by air raids in 1944, production resumed in 1947 but changes in global industry led to the closure of the mine in 1959, and the coking plant in 1977. Finally, the blast furnaces closed in 1985: 'The fires which had been roaring 24 hours a day, every day of the year, lighting up the sky ... were finally extinguished. Duisburg was sitting on a ruin.'[6] Visually, the new landscape of Duisburg Nord and Emscher Park is a mix of long views and poetic vistas with the immediacy of industrial structures. The effect is a balanced asymmetry, not unlike that of neo-classical painting. Vertical iron structures and horizontal waterways set off enclosed gardens and open pathways. Decontamination leaves no visible traces, some areas being excavated and re-filled with soil, others to be naturally cleansed over many years by plant and tree growth in the remit of the Forestry Department, not the IBA.

The first sight, approaching from the tram stop at the edge of Duisburg, is a set of towers looming over newly regrown trees (Fig. 5.1).

Duisburg Nord offers a café, visitor centre, exhibition hall, open-air cinema, children's play areas, cycle tracks on old rail lines, and footpaths. On summer weekends there are heritage rail trips using old locomotives and carriages, and the station at which workers once began and ended a shift. A canal, previously an open sewer, is clean now. Rainwater is harvested for irrigation. Mayflies hover among yellow lilies and reeds in cooling tanks. More than a hundred plant species have re-emerged naturally, and several bird species have returned. For Annaliese Latz, of landscape architects Latz+Partner, who worked at Duisburg Nord, it is, 'a garden where we work or whose stillness and beauty we enjoy in contemplation.'[7]

I want to read Duisburg Nord not simply as an economic measure or provision of community benefits, but as an ambivalent post-industrial memorial. Architect Deborah Gans observes the wider re-greening of the Ruhr (10,000 hectares have been decontaminated) to say that this constructs a distributed landscape of many nodes and paths, opposed to the tendency to mono-functional zoning in modern urbanism. Gans summarises the outcome of the IBA's projects as a, 'loose sprawl of the post-industrial Ruhr laid upon a fractured and fissured ground of disused industry and allied worker settlements,' aiming for a new order so that,

Fig. 5.1 Duisburg Nord landscape park, seen from the tram stop. (Author's photograph)

'like the first industrialists,' the IBA has exploited the site but, now, 'for its latent urbanity as well as for its historical culture.'[8] But Gans uses the loaded term 'Lebensraum' (living space, used by the Nazi regime for east-ward expansion) for the past proliferation of industrial sites:

> The Emscher region is vast because industrialists saw the landscape as an open field, unchallenged and without impediment to free-ranging colonisa-tion – a Lebensraum. Structures were erected willy-nilly in the countryside, used until a resource was depleted ... and then moved. Industry blazed trails of contamination as it moved from south to north, mining, building and discarding. ... This shifting ground of employment subjected the region's initially mixed society of agriculture, steel and commerce to an increasingly focused idea of labour, of inhabitants who put their lives at the service of progress.[9]

Gans argues that the trails of previous contamination have become a, 'new green armature' connecting the Ruhr's 17 cities.[10] This does, as she says, denote a postmodern decentralised economy, and, at the same time, a masking of an inconvenient past of ruthless exploitation. I would add that it also masks a past of skilled, well-paid work which was not in itself destructive.

In a related way, cultural historian Kerstin Barndt writes that the re-greening of the Ruhr ignores histories of class and collective identity:

> The elevated markers of land art that now dot every other slag heap in the region can serve as indicators, for they inscribe the disappearance of labour into their scopic regime. The new vantage points invite adventurous climbers to rise above the reconstructed landscape and contemplate the view. This privileged and individualised vision is significant in the context of post-Fordist modernisation. The new landscape of affect ... symbolically enables visitors to rise above local history.[11]

Gans hints at this when she notes that the IBA reads 'quality' as 'aesthetic quality, quality of building design and construction ... [and] quality of life,' rather than as better social conditions.[12]

Barndt is more emphatic, reading the Ruhr's transformation as aesthetically driven erasure of one history in favour of another: a-historical narrative via architectural preservation. But perhaps this recycling of material form is part of capital's normal process. Philosopher Dylan Trigg argues that, 'places of labour and craft ... have been rendered superfluous' when, 'the cyclical nature of capitalism, whereby new industries suggest rational progress but only at the expense of destroying old industries ... [so] disorder and mutability are suppressed.'[13]

At Duisburg Nord, industrial structures are disconnected from their function, and from the histories of industry produced by either workers or the owners of capital, and from tension between those histories. Landscape architect Catherine Heatherington argues, however, that Latz+Partner enabled overlapping relationships to the site,

> Layers of history ... are made visible in the new landscape, creating narratives which can be read by the visitor in multiple ways. [This] ... is exemplified in Latz's use of the railway tracks; these stretch across the surrounding landscape towards the river Emscher and fan out in curves to the individual buildings ... Thus the view of the railways from the top of the blast furnaces, whilst alluding to the developing post-industrial landscape, situates the site

within the industrial history of the Ruhr, connecting it with the river Emscher, the waterway that brought raw materials to the area.[14]

Perhaps Heatherington's response reflects her knowledge of design and immersion in visual culture; but she also sees standardisation in, 'a common palette of materials – rust, steel, Cor-Ten, concrete, gablons – which … create a generic narrative,' becoming static memorial devices for an undisclosed past.[15] I agree; but to me the issue is the contrast of a seemingly timeless, singular, naturalised aesthetic to the specificities of plural human histories. That attempt at timeless art occurred, too, in the landscaped park (now seen as very eighteenth-century).

Barndt identifies a dialectic in post-industrial ruinscapes which relates to the landscaped park:

> Similar to British gardens from the eighteenth century, with their sweeping vistas and hidden temples, the new landscape in Duisburg features labyrinthine pathways, inviting visitors to discover surprising details and spectacular open views … [or] explore the gardens and playgrounds in the bunkers below. One step further down the terraced landscape of the industrial park, the theme of active recreation turns to contemplative, and the vista opens toward a garden scene among the ruins.[16]

Barndt adds that the site's postmodern eclecticism combines elements ranging from clipped hedges to wildflower gardens, and from industrial structures to recreational facilities. If there are commonalities between the ruinscape and the landscaped park they revolve around visual strategies such as contrasts of scale and the use of nodal points in vistas. Yet although the neo-classical statues and temples in a landscaped park emulated a classical past seen on the Grand Tour of European sites or in collections of pillaged antiquities, and their references were understood through a classical education which included Roman poets such as Ovid and Virgil, the democratised access of post-industrial ruinscapes offers people immediate access to a memory of modern industry which was material, not poetically imagined.

Looking back, the landscaped park arose in a context of the military maintenance of Empire, foreign wars and civil unrest at home, which it evaded through a mask of nature. It admitted that situation obliquely in allusions to other imperialisms and morbidities, on the basis that even the Roman Empire fell, and the British might, too, one day. But it presented this sign of temporality at a remove, and for an elite audience. In contrast, post-industrial ruinscapes are dislocations of a more recent history still remembered by many local visitors, under the cover of nature as a universally healing, reconciling principle.

This is the specifically postmodern character of ruinscapes. As Trigg writes, 'The twentieth-century lineage of preservation persists in the form of placeless sites which aspire toward autonomy and self-rule.'[17] Trigg cites Le Corbusier's self-contained and geometric aesthetic as a denial of impermanence. But while Le Corbusier looked to an aesthetic of solid blocks interspersed with open spaces, ruinscapes express their autonomy as rusting metal, falling towers, bits and pieces everywhere—as if to trash the modernist spatial dream—while retaining an uncontested if hollow grandeur.

Walking by the Emscher, a gantry stands by the now clean water; strollers enjoy the air and dogs investigate the river bank (Fig. 5.2). Watching them as I while away the day, I am aware of histories of environmental

Fig. 5.2 Emscher Park waterway. (Author's photograph)

damage but also, as said, the potential for workers' solidarity in modern industrial sites. I also see the aesthetic quality of the landscaping, its framing and gradations, its balanced asymmetries understated to the point of being just noticed. But this is because my background is in fine art, critical theory and contemporary culture. Aesthetic and social readings are not incompatible, of course, when the act of critical interpretation is a traverse along an axis between these polarities. But that reminds me how much I interpret what I see, that nothing is natural as such.

INDUSTRIAL NUANCES

Thinking of ambivalent histories leads me to Prigann's work at Gelsenkirchen, a redundant mining site (near his childhood home, and heavily bombed in 1942) where, commissioned by the IBA, he used fragments from demolitions to create a meandering trail through the newly regenerated woodlands. Large slabs were made into shelters supported by cement pillars. Elsewhere, smaller pieces, recognisable as part of a doorway or a decorative cornice, were used to make markers, steps and incidental points of interest along the way. The trail begins near the new science park and leads through the woods to where a view appears of the slag heap, preserved and stabilised. A spiral pathway leads to the top, where concrete slabs were used to form a sculpture, *Heaven's Ladder* (*Himmelstreppe*, 1997–2000), seeming to reach for the sky. Like the biblical Jacob's ladder it aspires to the impossibility of reaching heaven. But it is neither a folly nor an archaeological relic but the remnant of a legible past. Prigann writes,

> The goal: aesthetic transformation of the industrial wasteland into a fascinating landscape of experience. The idea: the beautiful unfolds in a landscape that reveals the traces of its depletion and destruction everywhere. ... The way: a long-term connection to the area. The process of aesthetic appropriation ... occurs on a step-by-step basis, without a previously set planning concept but with a vision. ... An archaeological field develops. The integration of succession [growth] areas and the maintaining of open spaces, the planting of wild dog roses and other plants follows. Routes and paths are constructed. ... On the peak rises a landmark.[18]

In conversations with Prigann in 2002, he called *Heaven's Ladder* a memorial to industry. He stressed that industry should not be dismissed because it provided the benefits of goods in daily use, and work and work-ers' solidarity (a view I have absorbed, as stated above); yet it was an envi-ronmentally destructive history, a degradation of the land which produced both local pollution and global heating.

Prigann called nature his collaborator, using succession growth in every project, and saw its work as nuanced, erasing his interventions yet not completely doing so, because drainage patterns retain traces of interven-tion for millennia. His own work will never be completed because succes-sion growth is integral to it, evolving (literally) wildly; at the same time, his projects have beginnings, development phases and adaptations, and endings in the sense of moving on. But the completion is at best contin-gent on succession growth, and has its own timescale. This *non-teleological* position contrasts with the IBA's aims, and with the generic palette of post-industrial ruinscapes elsewhere. Pieces of concrete and brickwork survive as scar tissue in a never-ending coming to terms, which appears a viable equilibrium between memorialising the good side of modern indus-try and admitting its negative impacts.

Prigann was aware of the mystique of ruins. At a redundant water puri-fication plant for the chemical industry in Marl, also in the Ruhr, he cre-ated earthworks and a pool within the complex, retaining partly demolished walls and pipes (Fig. 5.3). The site was useless when a rise in the water table caused by turning off the pumps in disused mines in turn caused a fracture of the purification plant's circuits. Repair was not economically viable, and industry was leaving the area. Now, industrial remains, banks of.earth, and reflections in the pool resemble a Romantic image of tempo-rality, a forlorn grandeur. Like the empires remembered in statues of which few people now know the names, industrial monuments speak of having fallen, or having lost their currency.

Walking into the site was for me, at a stretch of the imagination, like walking into the pylon hall of an Egyptian temple although, again, I felt this because I *had* been to such places and was looking for analogies within my own cultural vocabulary. The meanings of ruins are as personal, then, as they are paradoxical: they speak of endurance, are anthropomorphised in terms of fortitude and resilience, but are pathetic (evoking sorrow, or in art the feeling of transience). My lingering view of Prigann's work at Marl is as an articulation of continuity (the ruin remains) and rupture (it is bro-ken). In another way, I read ruinscapes generally as erasing histories of work and solidarity in everyday industrial towns while providing leisure resources for communities. Post-industrial ruinscapes may be devoid of

Fig. 5.3 Herman Prigann, earthworks at Marl. (Author's photograph)

the social meaning, and implicit human values, of their sites (sometimes exhibited in documentary form), but do they act as memorials?

War memorials subsume individual grief in an overarching narrative of national identity; and perhaps post-industrial ruinscapes subsume stories of industrial life in a meta-narrative of a new economy. But I think this is problematic due to industry's production of climate change, which may be why it is masked by the lushness of re-greening. But if readers complete the meaning of literary texts through their projections, associations, and allegiances, then so the publics who visit and use post-industrial ruinscapes extend the meanings of those sites in ways which are personal, but not only personal. That would be more than the usual run of war memorials allows.

Unsanctioned Use and Urban Exploration

While Duisburg Nord provides sanctioned leisure uses, in Britain and North America it is not uncommon for such sites to be abandoned. Then they are outside regulation, inviting illicit uses. Such sites are seen by planners, developers, local politicians, and residents as derelict, awaiting an economic upturn to be reconsidered for use; but they are also disconnected from the networks of work which gave them meaning. Being *allowed* into such sites denies the positive agency of reclamation. Geographer Tim Edensor writes of a desire for, 'surprise, contingency and misrule' in place of the, 'seamless order' of a social stability which never existed but symbolically dominates urban redevelopment.[19] He notes that factories relied on an efficient rhythm of labour, but that factories also housed informal choreographies in, 'the chatter of the workplace, the constant whir of machinery, the songs sung together and all the other elements of the habitual soundscape forged [into] a familiar backdrop of routine.'[20] This, he adds, is why the silence of a post-industrial ruinscape is so captivating. And traces—old notices, bits of machine, the litter of an uncleared site—foster 'uncanny reflections from who-knows-where,' which reveal memory's intertextuality, the complexity of place as opposed to space, 'haunted by the social.'[21] Edensor continues,

> Whilst ruins can throw up the utterly strange and the very familiar, the uncanniness of that which is frightening and strange but simultaneously comforting and familiar ... provokes nebulous memories, for to confront such things is to encounter a radical otherness which is also part of ourselves. Partly, this is connected with the alterity of the past and the impossibility of reclaiming it whole ... [when] traces of our past selves leak out from a present in which we have tried to contain and encode the past. But it is also because ruins are rampantly haunted by a horde of absent presences ... [which] are ultimately evasive and elusive.[22]

I want to make a speculative leap from Edensor's geographic imagination to the idea of misrule—as in Carnival, or the Land of Cockaigne where a roast duck flies into the scholar's mouth (in a painting by Pieter Brueghel)[23]—as a politicised alterity.

The dominant order in this case is scientific rationalism, shaping entrepreneurial capitalism to reach a peak in the eighteenth century, mitigated by nineteenth-century reformism and the twentieth-century Welfare State, returning now in a total domination. But if scientific rationalism is a conflation of an evolution from primitive animality to a human autonomy with another from knowledge as finding-out to knowledge as power-over, then ruinscapes offer a glimpse of its fallibility. This is complex. Rationalism is freedom from the indifferent, unbending rule of mysterious Fate; a disenchanted world is a world freed from beguilement and superstition, which I welcome. But in Romanticism the wild is returned, fracturing the base of rationality in depictions of ferocious storms, wild seas, and volcanic eruptions (which inform depictions of the first, rural industrial sites), to say that any human intention is brittle and likely to be broken. So, ruinscapes play havoc with history, haunted yet familiar in the view of an interloper. Trigg writes, from post-industrial sites in North America,

> The double life of the ruin, as a shadow of its former being, now subverts that presence ... We can think in terms of the history of an absolute past, rational and unyielding. What emerges in the present ... is an event that does not belong in the present, and is not recognised in that present.[24]

Later, Trigg cites the English novelist Rose Macaulay, who wrote in the *Pleasure of Ruins,*

> Very soon, trees will be thrusting through the empty window sockets, the rosebay and fennel blossoming with the broken walls, the brambles tangling outside them. Very soon, the ruin will be engulfed, and the appropriate creatures will revel. Even ruins in city streets, if they are left alone, come, soon or late, to the same fate.[25]

For Trigg, this affirms the alignment of modern ruins with the decline of reason. But he also notes, 'The exhaustion of things outlives their physical demise.'[26] This indicates a spectre of purposefulness in a site of purposelessness. Trigg argues from his reading of the Kantian sublime that, 'the native context of the post-industrial sublime is not the halo of ascent but the flickering resonance of descent and gravity.'[27] For Macaulay, in *The World my Wilderness*, written after her experience as an ambulance driver in the blitz, and loss of her flat and all its contents from a direct hit, the ruinscapes of post-blitz London incite insubordination.

Macaulay's protagonist, Barbary (the name is indicative) at first lives an unruly life in the south of France during the Vichy state. She runs with the *maquis* (the Resistance) but at a crucial point, after her mother's lover has been assassinated as a collaborator, she is sent to London to live with her respectable father and his equally respectable second wife. Barbary responds to their imperative of good behaviour by running in the bomb-sites (the wilderness of the novel's title) with her cousin Raoul. Wandering the bombsites, improvising a living by making sketches to sell to tourists (in the immediate post-war years), she makes a home of sorts in a bombed building. They steal bicycles, acquire ration books by dubious means, and mix with marginal people. The police are the gestapo. No-one can be trusted. Only when Barbary has an accident does she find an enforced stasis. At the end she becomes an art student, but by then the bombsites are flowering: fireweed, pink rose-bay, red campion, yellow charlock, brambles, bindweed, thorn-apple, thistle, and vetch all proliferate of their own accord, untended and rampant.[28] Small gardens have been planted, signs of human re-ordering while the will to recovery strives against, 'the drifting wilderness to halt and tame it.'[29] This is a reconciliation between wildness and ordering, a recognition that destructive gestures are liberating but ephemeral; and that the memory of the wilderness lingers, refracting the present.

I wonder if urban explorers create a similarly refracted image. Geographer Oli Mould reads urban exploring as a, 'subversion of hegemonic urban control,' differentiated from, and beyond, the market's appropriation of urban subcultures.[30] Street art, for instance, is the market's appropriation of graffiti for urban art collectors. Skateboarding is more difficult to take over, being an ephemeral action; although local authorities provide skateboard parks, being sanctioned in this way removes the excitement of illegality.[31] Skateboarding in the undercoft of London's South Bank Centre has become part of the tourist route, regularly photographed, but retains its autonomy to an extent.[32]

Nonetheless, urban exploring differs from these subcultures in being carried out beyond the public gaze, in secret, both in old structures such as the catacombs of Paris or Rome, or disused nuclear bunkers, or in structures currently in use such as storm drains and utility tunnels. Urban explorers aim to leave no traces, but take and circulate photographs of interiors and views from heights. The record of their illicit action has not, as far as I know, produced a market for these images as aesthetic objects of increasing price and conspicuous consumption.

Mould's agenda is to subvert the Creative City myth, in which a new art museum or a new piazza is a means to increase property values by visibly re-coding a district. I agree with his argument (and have made my own elsewhere).[33] Mould sees urban exploration as exemplifying urban subversion, as an activity which is outside the market's reach, necessarily an alterity. He cites urban explorer Bradley Garrett's climbing of the Shard (London's tallest building) in 2012, using internal stairs and finally a crane on the summit[34] Garrett calls this place-hacking, which aligns the activity to computer hacking, which similarly plays with the borders of legality.

Mould recalls his own exploration of the Woolworths building, New York in 2013 with a small party of urban explorers:

> So, dressed like we were supposed to be there (in a suit …) we sweet-talked our way past security on the ground floor … and took an elevator to the 22nd floor. Trying to look nonchalant and inconspicuous, I felt anything but, my eyes flitting between the strangers I had just met …and the other strangers in the lift who presumably had no idea (or care) about what we were doing.[35]

What they were doing was going to the upper floors, intended for condominium conversion but left semi-derelict by the financial services crash in 2008. The explorers find a door, go up a stairwell into thirty storeys of loose electrical wires, exposed girders, piles of plasterboard and rubble, eventually reaching the top floors where another door opens onto a view over the city. Mould rationalises the experience as taking possession of a view otherwise denied:

> By operating at the boundaries of the city's institutions, we had glimpsed a … view of Manhattan from a vantage point that few would ever get to see … created new functions, subjectivities, experiences and social connections that would not have been achievable via a more official means.[36]

Such acts cannot be officially sanctioned. Perhaps they are interesting but not much use in radical social transformation or political struggle. But these actions are publicised on social media, and in that way contribute to social consciousness. They are ephemeral and unlicensed, and may occupy a place on a discursive axis between the ephemeral-illegal, and more organised occupations such as Occupy in 2011–12 and Extinction Rebellion in 2019–20, when representational politics is inadequate for current political needs.

Besides, ephemerality does not prevent transformation; it is a matter of who (rather than what) is transformed. For those taking part in Extinction Rebellion—I speculate, being a spectator for a day—the act is ephemeral but the consciousness of being there among others of like mind is inherently transformative, extending the horizon of the possible, opening a radical imaginary without which social processes reproduce past constraints. In the next chapter, I juxtapose community enterprise to a vogue for new public spaces in urban redevelopment. Here I give the last word to architect Rahul Mehrotra, on ruins and contemporary cities in India,

> The ruins of modernity are a fascinating intersection where the static city, encoded in architecture ... becomes modernity's ruins and creates a moment for contemplation. The ruins are positioned between their former newness, as symbols of optimism, and their ultimate implosion as they are engulfed by a landscape that they set out to re-make ... The modernist ruin dissolves its utopian project and becomes a monument that symbolises our historical trajectory by fabricating multiple dialogues with its context.[37]

NOTES

1. Sennett, R. (1998) *The Corrosion of Character: The Personal Consequences of Work in the New Capitalism*, New York, Norton.
2. Lodziak, C. (2002) *The Myth of Consumerism*, London, Pluto.
3. Kremmer, M. and Schupp, F., [c. 1932] (2011) quoted in Kift, R., *Tour the Ruhr*, Essen, Klartext, p. 125 [no source given].
4. Kift, *Tour the Ruhr*, p. 12.
5. IBA (1999) '1989–1999 IBA Emscher Park: A future for an industrial region,' www.open-iba.de/en/gesichte/1989-1999-iba-emscher-park [accessed 25 September 2018].
6. Kift, *Tour the Ruhr*, p. 75.
7. Latz, A. (2010) 'Regenerative Landscape – Remediating Places,' in Tilder, L. and Blostein, B., eds., *Design Ecologies: Essays on the Nature of Design*, New York, Princeton Architectural Press, p. 189.
8. Gans, D. (2004) 'The Sky Above and the Ground Below,' *Architectural Design*, 74, 2, p. 51.

9. Gans, 'The Sky Above and the Ground Below,' pp. 51–52.
10. Gans, 'The Sky Above and the Ground Below,' p. 52.
11. Barndt, K. (2010) 'Memory Traces of an Abandoned Set of Futures: Industrial Ruins in the Post-Industrial Landscape of Germany,' Hell, J. and Schönle, A., *Ruins of Modernity*, Durham (NC), Duke University Press, pp. 277–278.
12. Gans, 'The Sky Above and the Ground Below,' p. 53.
13. Trigg, D. (2009) *The Aesthetics of Decay: Nothingness, Nostalgia, and the Absence of Reason*, New York, Peter Lang, pp. 120–121.
14. Heatherington, C. (2012) 'Buried Narratives,' Jorgensen and Keenan, *Urban Wildscapes*, London, Routledge, p. 180, citing Kamvasinou, K., 2006, 'Vague Parks: the politics of late twentieth-century urban landscapes,' *Architectural Research Quarterly*, 10, 3–4, pp. 255–262.
15. Heatherington, 'Buried Narratives,' p. 183.
16. Barndt, 'Memory Traces of an Abandoned Set of Futures,' pp. 279–280.
17. Trigg, *The Aesthetics of Decay*, p. 207.
18. Prigann, H. (2004) 'Rheinelbe Sculpture Woo, 1997–2000, Gelsenkirchen,' Strelow, H., ed., *Ecological Aesthetics: Art in Environmental Design: Theory and Practice*, Basel, Birkhauser, p. 132.
19. Edensor, T. (2005) *Industrial Ruins: Space, Aesthetics and Materiality*, Oxford, Berg, p. 55.
20. Edensor, *Industrial Ruins*, p. 149.
21. Edensor, *Industrial Ruins*, p. 151.
22. Edensor, *Industrial Ruins*, p. 152.
23. Bloch, E. [1959] (1986) *The Principle of Hope*, II, Cambridge (MA), MIT, p. 813.
24. Trigg, *The Aesthetics of Decay*, p. 133.
25. Macaulay, R. (1977) *The Pleasure of Ruins*, London, Thames and Hudson, p. 237, quoted in Trigg, *The Aesthetics of Decay*, p. 137.
26. Trigg, *The Aesthetics of Decay*, p. 138.
27. Trigg, *The Aesthetics of Decay*, p. 148.
28. Macaulay, R. [1950] (1958) *The World My Wilderness*, Harmondsworth, Penguin, p. 187 [paraphrased].
29. Macaulay, *The World My Wilderness*, p. 187.
30. Mould, O. (2017) *Urban Subversion and the Creative City*, London, Routledge, p. 115.
31. Borden, I (2001) *Skateboarding, Space and the City*, Oxford, Berg.
32. Mould, *Urban Subversion*, p. 142 [image].
33. Miles, M. (2015) *Limits to Culture*, London, Pluto.
34. Mould, Urban Subversion, p. 113; www.placehacking.co.uk/2012/04/07/climbing-shard-glass [accessed 29 February 2020].
35. Mould, *Urban Subversion*, p. 186.

36. Mould, *Urban Subversion*, pp. 187–188.
37. Mehrotra, R., (2010), 'Simultaneous Modernity: Negotiations and resistances in Urban India,' Hell, J. and Schönle, A., *Ruins of Modernity*, p. 248.

CHAPTER 6

An Urban Revolution?

Abstract This chapter begins in The Tide, a new public space in London's Greenwich peninsular. It questions the myth that new public spaces foster democracy, arguing that they aim to increase property values. This is the latest means by which the potential of urban living is undermined and reduced to a predetermined and uncreative routine. The chapter reiterates aspects of previous chapters to move to an alternative approach in the social housing of Red Vienna in the 1920s, and the conditions of proximity, diversity, mobility, and agency in localised housing and community enterprise projects today.

Keywords Public space • Creative city • Radical democracy
• Red Vienna • Social housing

The Tide is a new public space on the Greenwich peninsular. It begins with a flight of steps rising from the plaza opposite the exit from the London Underground station. Hoardings advertise the outlet shops in the Millennium Dome (O_2 Arena) on one side; on the other side are steel-and-cladding boxes and towers for a higher education institution, residential spaces, and leisure enterprises. Designer-trees in tubs lend a green note; neat signs promote exercise and well-being on the way to a view of the Thames (Fig. 6.1).

© The Author(s) 2021
M. Miles, *Paradoxical Urbanism*,
https://doi.org/10.1007/978-981-15-6341-6_6

Fig. 6.1 The Tide, North Greenwich, London. (Author's photograph)

The Tide hovers over an open public space, with no obvious reason to add to it. Architecture critic Oliver Wainwright sees it as a, 'souped-up graveyard of novelty trinkets ... to elevate the value of a steroidal development of luxury apartments.'[1] I agree. Wainwright notes the precedent of the High Line, a disused rail viaduct in Manhattan which has been landscaped and opened to public access. But while the Highline is so popular there can be queues to get on it, The Tide looks redundant in advance, its neat little trees looking as if just bought from an up-market garden centre, which they probably have been. There are the now obligatory public sculptures to say this is high-rent space (aligned to the blue-chip art market) but Wainwright continues,

> Standing on the elevated deck, looking out over a jumble of vents and service hatches, it's difficult to work out how anyone thought this was a good idea. It has no purpose ... apart from providing a slightly different perspective on the surrounding carnage. A flight of raked steps suggests an area where performances might take place, while another section swoops down to form a protective bowl around a café terrace, but the point of being raised ... over an area that is already pedestrianized remains a mystery.[2]

The Tide is public in designation only, and exemplifies the new piazzas which have become as obligatory and predictable as public art in redevelopment schemes. There are well-used, well-conceived public spaces—such as Centenary Square in Birmingham, with brick paving designed by artist Tess Jaray,[3] which is on a pre-existing pedestrian route in the city centre—but The Tide is formulaic and unnecessary, connecting nothing with nothing (to paraphrase T. S. Eliot's *The Waste Land*, 1922).

The proliferation of such piazzas in British cities follows the report of an Urban Taskforce led by architect Richard Rogers, which considered economic growth and citizenship (the latter after introduction of a category of social exclusion by the Blair regime).[4] Rogers advocated better housing and play spaces; and saw new public spaces as likely to foster participation in civic life as a means to address social exclusion. Advocating better housing echoes the liberal reformism of the nineteenth century, seeing improvements in the material conditions of the lower classes as improving their behaviour. But the piazzas are decorative, low-cost visible signs of renewal which do not touch underlying problems such as multiple deprivation and lack of social mobility. The report is bland. Its objectives are undemonstrable. For instance,

> We want our towns, cities and suburbs to be places for people: places that are designed, built and maintained on the principle people come first. They should contribute to the quality of life and encourage healthy and sustainable lifestyles. They should be places in which we want to live, work, bring up our children, and spend our leisure time.[5]

I agree with Owen Hatherley when he comments, 'The Urban Renaissance was the very definition of good ideas badly thought out ... pointless piazzas with attendant branches of Costa Coffee.'[6] Other coffee chains exist. What else?

ANOTHER KIND OF CITY?

At the book's outset I identified four elements which I took as key to city living, all aligned to an emerging or imagined public sphere of social determination. They are:

- proximity—the co-presence of many people in high-density occupation;
- diversity—being amid people from different genders, ages, races, and other social categories, on a presumption of equality;
- mobility—social as well as geographical;
- agency—the civic value of co-determination.

I cite local initiatives at the end of this chapter which indicate the viability of practical means to enact values which might produce these conditions. These are local, small-scale projects, but which together indicate a new society taking form within the old. But before looking at those cases I want to briefly reiterate what I have taken as obstacles to the potential for city living: a rural myth, and an instrumentalism embedded in modernist urbanism. Both are ingrained in modernist urbanism, and its postmodern successor signified by redevelopment schemes which clear local populations for vacant sites to serve global capital's imperative of wealth accumulation (and its other side, impoverishment of the majority). If city air makes people free (Chap. 1) the question is who (which people) as well as how.

THE RURAL IN THE CITY

Suburbs had trees, and houses with gardens: bits of countryside on an urban periphery (but less peripheral than sites which had a history of outsider status, like the Zone in Paris).[7] They fostered, Richard Sennett shows, an enclosed, unreal sense of community based on white, middle-class notions of identity, purity, and family values (Chap. 2).[8] The initial impetus to re-invent rurality was a reaction against industrialisation but its persistence requires further consideration. For example, the Clachan, a whole Scottish Highland village, was exhibited at the 1938 Empire Exhibition at Bellahouston Park, Glasgow. The Exhibition aimed to reassure visitors from home and abroad that the British Empire prevailed (approaching war with Nazi Germany) and to celebrate modern design. Many pavilions were in Art Deco style, including the Palaces of Engineering and Industry. The

Fig. 6.2 The Clachan (Highland Village), 1938 Empire Exhibition, Glasgow. (Postcard, author's collection)

Tower of Empire, designed by Thomas Tait, was three hundred feet high, sited on a hill, with (very modern) cantilevered sections.

The Clachan (Fig. 6.2) included cottages, shops, a castle rampart, and a small lake. Tait's tower loomed over it, making the Clachan seem regressive, a rural presence in an urban scene, and one which ignored the real hardships of life in a Highland village but represented some kind of continuity in a world recovering from economic depression, facing war.

I have a set of postcards of it. One shows a boy in Highland costume leaning against a tree as a girl sits on the grass (below him) by a lake. A rowing boat rests on the shore. Visitors walk a sylvan path. Another card shows a thatched cottage, and a whitewashed building, possibly a shop. A third depicts a rural post office—a single-storey, thatched stone-walled building—with spectators (all wearing hats). In a fourth, a group of people on the bridge are having a picnic (also wearing hats). The castle has one round tower, one square.

I can imagine what it was like from visiting the rural life museum at St Fagans, Cardiff, part of the National Museum of Wales. But St Fagans is educational, charting the social changes of a mining community in the changing interiors of a row of cottages, for instance. Like the Clachan, no-one lives there, but no-one pretends it is other than an architectural reserve. Looking at the postcards and the well-dressed visitors I wonder if, for them, the Clachan was like going to the cinema, an escape from the world outside; or like a charabanc trip to the countryside as escape from the routines of toil. Well ... it was more than eighty years ago so I cannot ask them.

The only structure from the Glasgow exhibition which remains on site is the Palace of Art designed by Lancelot Ross. The South African pavilion was re-used as a chemical company's canteen; and the Palace of Engineering became part of an aircraft factory. Then there was the war, and then post-war rebuilding and new city centres such as Plymouth (Chap. 4). The only trace of a rural idyll in Plymouth is the spine of landscaping in Armada Way, which Patrick Abercrombie specified should introduce the delights of the gardener's art to citizens. But Abercrombie did not put a village there, with living inmates. What he did, however, was use a conventional top-down method of planning.

The Inscription of the City

Henri Lefebvre is insistent that lived space, where everyday life takes place, is in a dialectic relation to planned space within a society's overall spatial practice.[9] Although lived space is marginal it returns incidentally. If I sit on a public step, or project personal associations onto a space, I reclaim it. Being produced through acts of occupation, lived space can be enabled but not designed as such. Instead, it instantiates spontaneous self-organisation, and lasts but a moment.

The co-presence of lived and planned space, one by design, the other a process, underpins insight as to how regressive myths and top-down inscription can be refuted. For example, the open space outside the Museum of Contemporary Art in Barcelona (MACBA) is used by dog walkers in the morning, skateboarders a bit later, then motorbike riders. Nearby, the New Ramblas, made by demolishing five blocks of nineteenth-century apartment buildings, has a range of users and uses. It was intended to sanitise the old red light district (*barrio xino*) and denote a cultural quarter as engine of regeneration. This was unsuccessful, and several of the new art galleries closed. But, from a longitudinal study of the site, sociologist Monica Degen writes that diverse publics have created an emotional topography. Long-term elderly residents, young incomers, homeless people, migrants and tourists co-negotiate its use in unspoken but real ways.[10] There are visible changes but these are accommodated in a longer-term social process affirming the social ownership of the site: a viable proximity and diversity.

This was unplanned but not inarticulate. The danger is that the unplanned is invisible, and ignored, causing damage when local cultures are unrecognised in the developer's gaze; or are purposefully erased as obstacles to redevelopment. For example, the last hole-in-the-wall oatcake shop in Stoke on Trent in the English West Midlands was closed in 2012. Oat cakes are a traditional local food, cooked on site, sold directly through a window. The shop's owner, Glen Fowler, said, 'people who come here are not just customers, they're friends.'[11] The city's Member of Parliament, Tristan Hunt (now a museum director) lamented, 'we've knocked down too much of the city's fabric.'[12] He could have added that small shops pay small rents and serve small publics, but keep the money in the local economy. I was too late to see it; but noted that the local authority had offered houses for £1, with a refurbishment loan thrown in, to stem a shrinking population.

I do not know what the solution is for Stoke's decline, and distrust the notion that there are solutions: predictions based on selective interpretations of conditions, assuming that plan A means outcome B. This is countered by chaos theory—small shifts in conditions mean large changes in outcome for a given intervention—and ignores the complexities of emotional ownership and tacit knowledge. As geographer Jennifer Robinson writes from research in Manchester in the 1980s, people belong to multiple, overlapping or contesting communities of interest, hence, 'the character of urban social relations is diverse and changing.'[13] Urban redevelopment schemes, in contrast, even if well-designed visually, rely on a standardising image.

For instance, Hafencity, the redeveloped port area of Hamburg, has many public spaces with waterside views. It is mixed use, with a concert hall, maritime museum, theatre, and science centre. All very nice, but Hafencity exists because it also houses the Hanseatic Trade Centre, a cruise line hub, corporate offices, apartments, and sixty thousand square metres of retail space. Oddly, ElbElysium retirement home sits amid all this, offering waterside views for old people who, 'want to lead an active life in an urban environment.'[14] The Elysium (or Elysian Fields) of classical mythology was populated by the dead, not those awaiting death, but the name may be apt because they were the privileged dead who enjoyed music and athletics, as in real life. Admission was by invitation of the gods. In contrast, anyone can go to the six 'quarters' of Hafencity, of which Überseequartier is, 'a truly Archimedean point,' with, 'an open air shopping experience with pedestrian zones and plenty of public spaces.'[15]

Everything is an experience now, or a solution. And today's elites are global flâneurs exploring the wild lands of consumerism. Architect Krzysztof Nawratek writes of 'fluid and nomadic capital' which encroaches on spaces previously, 'liberated from its influences.'[16] The appropriation of public space illustrates this while ordinary life takes place in domestic and transitional spaces. Design can contribute to it but not in grandiose projects, more below the radar of recognition, as in social housing schemes.

If the clearance of the Heygate Estate (Chap. 1) shows that redevelopment's imperative is a void Cartesian space, its opposite requires dealing with a fear of disorder – unruly publics – as in the suburb, and in the deliberations of the Congress of Modern Architecture (CIAM, Chap. 4). This fear, which is a fear of mortality, led to instrumentalist planning and design in an effort to give a guise of permanence. Ends justified means, and this embedded way of apprehending reality became the norm. The language of planning and architectural design, separated from the work of building and multi-sensual acts of occupation, precludes the co-production of space, which might be a point of departure for an alternative way of making cities. Looking at the late-twentieth century, geographer Garry Bridge writes,

> Thus the potential of the city to provide the basis for a self-constituting public, beyond the partialities of associative civic life, were lost. In these conditions urban public space, even where it was generally accessible and a potentially open space of unpredictable encounter, served no broader function in terms of the constitution of a public realm.[17]

This draws out the mythicised status of public space. It does not mean there is no need for planning, but planning needs to depart from hermetic (Cartesian) systems. For sociologist Elizabeth Wilson, whom I cited in Chap. 1 and cite again here, 'What needs to change is the ultimate purpose of planning.'[18] That involves moving from a role of policing and social engineering to one which hands over agency to dwellers as co-producers of cities. When Wilson writes of, 'the deep ambivalence towards city life that has characterised modern western culture,'[19] she invites a reconsideration of what constitutes city life.

I have argued that such ambivalence rests in the rural myth; and in the instrumentalism of modernist urban planning and design. But, in a book about paradoxical currents, I accept a need for intention, for instance for a better world, but not for a blueprint or master plan. Design which is *produced by* a public aware of its agency may seem a dream today, but I think there was a glimpse of this in the social housing projects of 1920s Vienna. These were directed by the municipality and employed architectural knowledge, but were politicised, reasserting the right to the city of the working classes. It was a start.

DOMESTIC SPACE AS URBAN REVOLUTION

In Vienna in the 1920s—Red Vienna, when the city was governed by the Social Democratic and Workers' Party (*Sozial-demokratische Arbeiterpartei*, SDAP)—domestic spaces changed the image of a modern city by changing *for whom* it was built. Architectural historian Helen Meller notes that pre-1914 Vienna was, 'the cultural cradle of modernity,'[20] and a rapidly growing industrial centre the operatives of which were migrants from the provinces, living in shared rooms, using beds in shifts to fit day or night working. With defeat in 1918 and the consequent breaking up of the old Empire, the Social Democrats gained control in Vienna, retaining it even when losing it elsewhere in Austria. They modernised public services and transport, and took social housing to a new level of comfort and priority.

Since the 1870s, when the Ringstrasse was built following demolition of the city's medieval walls, housing had occupied city-centre sites. It was mainly designed in modern styles but for bourgeois apartments while workers commuted from the periphery. Some new social housing bocks were built in the 1890s and 1900s, however, among them the Metzleinstaler project, which architectural historian Liane Lefaivre describes as becoming, 'the prototype for all further council housing' in Vienna.[21] Metzleinstaler had running water, indoor toilets, and living rooms with a kitchenette and a stove. Its designer, Hubert Gessner, was a friend of Victor Adler, co-founder of the SDAP. Gessner also designed Reumann Hof (1924–26) with 392 apartments, on a street called Gürtell. The magazine *Die Unzufriedne* (*The Dissatisfied*) illustrated Reumann Hof under a banner, '*Die Ringstrasse des Proletariats*' (the Ringstrasse of the working class).[22]

Meller notes a fall in the birth rate across social classes after 1918, and economic decline. She writes,

> It was in this context that the SDAP set out on the most spectacular element of its social policy in the 1920s: the building of working-class homes. It was a policy which fitted theory, possibility (since taxes could be levied directly to pay for it), and practice (as the municipal bureaucracy could handle it). It was also an electoral winner as housing for the workers ... became a widely supported issue.[23]

Proposals were published in 1927 as *Das Neue Wien* (*The New Vienna*), including housing in blocks with garden courtyards on a much larger scale than earlier projects such as Reumann Hof. Most notably, Karl Marx Hof is a row of blocks more than a kilometre long, with 1382 apartments, generously sized for the time (thirty to sixty square metres, meeting standards at the Bauhaus and CIAM) (Fig. 6.3). Domestic space was matched by shared spaces, such as a laundry, playgrounds, a library, a primary school, a doctor's surgery, and business premises. Today there is a small museum. A metro station provides a fast link to the city centre.

Karl Marx Hof was designed by Karl Ehn, and built between 1927 and 1930. The design style is modernist, demarcated in red and cream and nodding to Art Deco. But more importantly, Karl Marx Hof is a statement that working-class people should have good housing, not as an inducement to good behaviour but as a right. Lefaivre comments that its scale meant that Karl Marx Hof 'could sustain health and education programmes, and parades, festivals and sporting events designed to create a new working class.'[24] It aided the growth of working-class culture as well as comfort, following, as Lefaivre says, a political agenda:

> One of the features that set Red Vienna's housing programmes apart from those of other cities ... is the sense of civic duty which led major cultural and scientific figures to contribute to the common good. The Social Democratic Party's two founders, Victor Adler and Otto Bauer, who abandoned prestigious professional life to serve the municipality, set the tone.[25]

Sigmund Feud established a free clinic, the Ambulatorium, for working-class people in 1922. And in Vienna's literary circles, Karl Krauss gave free public readings on political and cultural figures including Rosa Luxemburg and Bertold Brecht.

Fig. 6.3 Karl Marx Hof, Vienna. (Author's photograph)

But politics in Red Vienna emphasised the provision of domestic spaces, not grand public buildings; and provision of intellectual debate through the Social Democrat newspaper *Arbeiter Zeitung* (financed by a legacy from physicist Ernst Mach). There was a Workers' Symphony Orchestra, and Workers' Olympic Games (*Arbeiter Olympiade*) in 1927. From its cultural programmes, Lefaivre notes, the SPAD made gains in the 1927 municipal elections, while, 'The concert commemorating the tenth anniversary of the republic placed the span of Western music at the service of the proletariat.'[26] All this ended in 1933 with the rise of Fascism, but the buildings remain, a living monument to the just city.

The project has been resumed, as in Sonnwendviertel (Midsummer neighbourhood), near the Central Station, which includes five thousand apartments aiming for social as well as environmental sustainability, with a diversity of housing units and spatial uses. Among the architects selected,

Klaus Kada's design is characterised, Lefaivre says, by, 'an exceptional concept of community and open space,' with a library and meeting room, and a mezzanine which is open to use as a gallery, theatre or cinema, plus a communal kitchen, climbing wall, greenhouse, youth club and skateboard ramp, pool, café, and observation gallery.[27]

Socially progressive housing projects are increasingly common in Britain, too, both municipal and self-build. A case of the former is Goldsmith Street, Norwich (2019), a hundred homes in terraces built to minimal carbon standards using the German system *Passivhaus*, close to the city centre. The scheme is dense low-rise. Each apartment has a street door, which eliminates the common parts of blocks, while communal activity occurs in a play area, adjoining garden spaces, and a back alleyway. Another case is Lilac Housing in Leeds, an eco- as well as co-housing scheme initiated by a group of dwellers including geographer Paul Chatterton from Leeds University. It includes twenty dwellings, a shared allotment, bike sheds, and a common house where residents meet and can eat (to which the mail is delivered). Designed by White Design, the houses use straw-bales to fill lime-surfaced pre-fabricated timber cells, each super-insulated and airtight, with mechanical ventilation heat recovery units and solar water heaters. The group maintains a set of shared values with corresponding responsibilities for child care, cooking and other social tasks. The site operates a mutual ownership scheme by which each member pays 35% of their income to it. Asked what is wrong with standard development, Chatterton responds,

> It doesn't seem to be able to develop or deliver low-impact housing that represents the step change in carbon reduction we need. ... It doesn't really allow us to lock in affordability and perpetuity for future generations ... [and] it doesn't seem to have an ability to build the meaningful sorts of relationships on which communities are based. It just perpetuates really corrosive individualism ... So put these things together and you've got an incredible dysfunctionality. ... We really need some quite hot-footed experimentation to respond to that.[28]

Yes. To me this exemplifies a process in which values are enacted, means in themselves.

A RIGHT TO THE CITY

Projects such as Lilac Co-housing reclaim a right to the city, to borrow Lefebvre's term. By it he meant a refusal of, 'the so-called society of consumption,' and dominance of Cartesian space.[29] He looks for what he calls transduction and experimental Utopia. Transduction is the identification of a possible objective based on the problematics of reality assuming, 'feed-back between the conceptual framework used and empirical observations,' and leading to, 'spontaneous mental operations by the planner, the architect, the sociologist, the politician and the philosopher.'[30] Experimental Utopia is the study of material consequences, and an interrogation of the criteria and rhythms of daily life, in a search for happiness. It refuses to privilege the individual elements of the city—structure, function and form—over their co-production and inter-relatedness, or over sub-systems of knowledge: 'living here or there involves the reception, adoption and transmission of such a system,' while dwellers know a great deal about how and where they dwell.[31] Lefebvre continues,

> Architects seem to have established and dogmatised an ensemble of significations, as such poorly developed and variously labelled as function, form, structure, or, rather, functionalism, formalism, and structuralism. They elaborate them not from the significations perceived and lived by those who inhabit, but from their interpretation of inhabiting. It is graphic and visual, tending towards metalanguage.[32]

Since translation of Lefebvre's *The Production of Space* in 1991, such critiques are familiar in Anglophone discourses; yet inter-disciplinarity and tacit knowledge are still marginalised by dominant systems in professional practice and academic work while for more obvious reasons being seen by developers as anathema.

'The Right to the City' reflects Lefebvre's idea that urban vitality involves spontaneity. It also echoes his idea of moments.[33] That is, anyone may experience, anywhere and anywhen, a sudden realisation of actuality (an *I see* instant). Moments are ephemeral yet lingering and transformative. This is simple to say but complex. To me, one implication is the importance of direct action, and the solidarity of people gathered in a common purpose and culture, the moment being awareness of this togetherness, which may be more significant than efforts to change state policy. Being there, that is, is the moment, and enough at the time. It has the same relation to representational politics as lived space has to planned space. It directly enacts alternative values rather than signposting them.

Lefebvre argues, however, that both planning strategies and dwelling tactics are necessary, in a relation not unlike that of conceived and lived space. So, coincidentally but not unlike Red Vienna, urban strategy, 'cannot but depend on the presence and action of the working class, the only one able to put an end to the segregation directed essentially against it.'[34] In 1968, Lefebvre saw workers' strikes and factory occupations as the real insurrection, arguing against Herbert Marcuse's faith in a class of students and young technocrats. Looking back, I note André Gorz' book *Farewell to the Working Class* (1980),[35] and the argument that the working class is colonised by consumerism, enthralled in a wheel of toil and compensatory leisure.[36] But in 1968, memories of the Popular Front of 1934 (which also organised factory occupations) instantiated the solidarity of organised labour. In both cases these were new societies within, interrupting, the old.

CHANGING CITIES

Today, redevelopment schemes render cities images of global market forces and corporate power, using aggressive, instrumental methods such as urban clearance. To varying extents, nation-states and municipalities are complicit, only rarely departing from a neoliberal script. An alternative requires an extension of the public imaginary—the horizon of the possible—and practical projects demonstrating the viability of dreams. Imagination of a better world is the first step; the next includes housing projects such as those cited above, and direct action, from anti-roads protest in the 1990s to Occupy in 2011–12 and Extinction Rebellion in 2019–20, as continuous, contingent, living articulation of empowerment.

I end by citing two projects in Anfield, Liverpool, where nineteenth-century terraces were acquired for demolition in a scheme for a new football stadium but left empty, abandoned as an inner-city wasteland. This is where to find Homebaked now, a community bakery, and Kitty's Laundrette.

Homebaked planned to open in 2013 with a grant from the Social Investment Bank, just as it was announced that the streets around the Liverpool Football Club stadium were scripted for redevelopment. But community action led to the bakery's realisation, and the retention of one street of housing for refurbishment. Today the bakery is known for the pies people buy to take to football matches, and for fresh bread at afford-able prices. In its own words,

> We are loved by our community and match day visitors to Anfield … Our bakery is a community space for local residents and visitors to the area. It is a place for people to meet, share stories, celebrate and develop ideas for our neighbourhood together. We employ and train local people and pay a living wage. Today we are the only producing business on our high street.[37]

Homebaked is regenerating the street brick by brick on a model of co-ownership, using a community land trust, with profits from the bakery—eight hundred pies sold on the day of the Premier League final in 2019—re-circulated in the local economy.

Kitty's laundrette is round the corner, near the terrace which will be refurbished (working with architects Urbed). Opening in 2019, the laun-drette is named after Kitty Wilkinson, a local Irish immigrant who, in the cholera outbreak of 1832, invited neighbours to wash their clothes and bedding in her house as a means to fight the disease. In 1842 she secured public funds for a wash house and public baths. Now the laundrette has recovered her history (Fig. 6.4).

Rachael O'Byrne, community engagement lead in the cooperative management of Kitty's, says that although the opening of the laundrette might seem a sudden change, 'in reality it has been decades in the making. People have lived and died in this community waiting for a grand regen-eration that never arrived.'[38] Most of them are on less than average earn-ings, so affordable laundry is helpful, as well as opening a new social space while, as artist and co-founder of Kitty's Grace Harrison remarks, 'most of the laundrettes left in Liverpool are all run by the same company,' and charge rates higher than Kitty's.[39] Harrison sees Kitty's as hosting arts and social activities for the neighbourhood, and as celebrating the contribu-tions of immigrants to communities. In its own words, Kitty's Laundrette says,

> We've got good coffee and great free Wi-Fi, so it's a nice place to chat, work and hang out while doing your self-service laundry and dry cleaning, or you can leave it with our lovely team … we're proud to continue the legacy of Kitty and the many working-class women who frequented the wash-houses of Liverpool.[40]

Fig. 6.4 Kitty's
Laundrette, Liverpool.
(Author's photograph)

This is imaginative but not escapist, rewriting past stories without con-structing a dominant narrative; and it articulates Lefebvre's Right to the City. He says, 'The *right to the city* is like a cry and a demand … [which] cannot be conceived of as a simple visiting right or as a return to tradi-tional cities. It can only be formulated as a transformed and renewed right to urban life.'[41] It includes a right to occupy space, to be visible and have voice, and to contest other claims. As Lefebvre writes in *The Urban Revolution*,

... there is nothing harmonious about the urban form and reality, for it also incorporates conflict, including class conflict. What is more, it can only be conceptualised in opposition to segregation, which attempts to resolve conflicts by separating the elements in space. This segregation produces a disaggregation of material and social life.[42]

This can be juxtaposed to planning academic Leonie Sandercock's argument for agonistic politics,

Negotiating peaceful intercultural coexistence, block by block, neighbourhood by neighbourhood, will become a central preoccupation of citizens as well as urban professionals and politicians. The right to the city is the right of all residents to presence throughout the city, the right to inhabit and appropriate public space, and the right to participate as an equal in public affairs, to be engaged in debating and designing the future of the city.[43]

I have reservations about block by block, which sounds like guerrilla warfare, in a tendency in urban writing to fetishise conflict rather than accommodate conflicting claims. Lefebvre's phrase 'the right to the city,' can also be criticised, as by planning academic Peter Marcuse, who argued for 'a' rather than 'the' right and city, and to recognise privileges as well as rights.[44] But the point remains that opposition to the dominant city needs alternative visions and practical examples. In this context, Marcuse argues for the method of commons planning to address issues of power,

A society without power need not be a society without order, just as the presence of power is not an adequate guarantee of order ... The philosophic issues are complex, but the distinction between power and authority is central. Power is the ability to have others do one's bidding, against their own interest and for the benefit of the holder of power ... Authority is likewise the ability to have others do one's bidding, not for the benefit of the holder of authority but for a collective benefit ... by rules agreed to collectively.[45]

This is power-to, which is never donated and only gained through contestation and self- or group-awareness.

If all this relates to small, localised, ephemeral projects, I see no problem there. Nor do I see the task as scaling up such initiatives. It is instead to realise that they collectively de-centralise social determination: a revolution by other means when mass revolt—October 1917 and Lisbon 1974 are encapsulated in history now—is unlikely. This is why direct action is vital, manifesting hope and renewing the vision of a society in which values of autonomy, mutuality, equality, and joy are directly enacted, not taken merely as signposts to a future which never comes but justifies anything.

If there is a New Jerusalem, it does not descend by magic on the last Day; it is produced, in lived spaces, in projects like Kitty's Laundrette and Homebaked, and in political imagination.

Notes

1. Wainwright, O. (2019) 'The short and winding road to nowhere,' *The Guardian*, 11 July, p. 10.
2. Wainwright, 'The short and winding road to nowhere,' p. 11.
3. www.tess.jaray.com/commissions [accessed 29 February 2020].
4. Department for Environment, Transport and the Regions (1999) *Towards an Urban Renaissance*, London DETR; DETR (2000) *Our Towns and Cities – The Future: Delivering an Urban Renaissance*, London, DETR.
5. DETR, *Our Towns and Cities – The Future*, section 4. 3.
6. Hatherley, O. (2010) *A Guide to the New Ruins of Great Britain*, London, Verso, pp. xv–xvi.
7. Sante, L. (2015) *The Other Paris: An Illustrated Journey Through a City's Poor and Bohemian Past*, London, Faber and Faber, pp. 53–70.
8. Sennett, R. (1970) *The Uses of Disorder*, New York, Norton.
9. Lefebvre, H. [1974] (1991) *The Production of Space*, Oxford, Blackwell, pp. 38–39.
10. Degen, M. M. (2018) 'Timescapes of Urban Change: The Temporalities of Regenerated Streets,' *Sociological Review*, 66, 5, pp. 1074–1092.
11. Fowler, G. (2012) quoted, Doward, J. and Simpson, D., 'Potteries mourn passing era as developers claim last oatcake shop,' *The Guardian*, 4 March, p. 13.
12. Hunt, T. (2012) quoted in Doward and Simpson, 'Potteries mourn the passing era,' p. 13.
13. Robinson, J. (2006) *Ordinary Cities: Between Modernity and Development, London*, Routledge, p. 57.
14. Meyhöfer, D. (2014) *Hafencity Hamburg Waterfront*, Hamburg, Junius Verlag, p. 64 [text in German and English].
15. Meyhöfer, *Hafencity*, p. 124.
16. Nawratek, K. (2019) *Total Urban Mobilisation: Ernst Jünger and the Post-Capitalist city*, Singapore, Palgrave Pivot, p. 31.
17. Bridge, G. (2005) *Reason in the City of Difference: Pragmatism, Communicative Action and Contemporary Urbanism*, London, Routledge, p. 86.

18. Wilson, E. (1991) *The Sphinx in the City*, Berkeley, University of California Press, p. 156.
19. Wilson, *The Sphinx in the City*, p. 157.
20. Meller, H (2001) *European Cities 1890–1930s: History, Culture and the Built Environment*, Chichester, Wiley, p. 77.
21. Lefaivre,, L. (2017) *Rebel Modernists: Viennese Architecture Since Otto Wagner*, London, Lund Humphries, p. 124.
22. Illustrated, Lefaivre, *Rebel Modernists*, p. 123, fig. 48, from *Die Unzufriedne*, 35, 30 August 1930.
23. Meller, *European Cities 1890–1930s*, p. 97.
24. Lefaivre, *Rebel Modernists*, p. 127.
25. Lefaivre, *Rebel Modernists*, p. 128.
26. Lefaivre, *Rebel Modernists*, p. 131.
27. Lefaivre, *Rebel Modernists*, p. 283.
28. https://transitionnetwork.org/news-and-biog/paul-chatterton-on-lilac-leeds-co-housing [accessed 6 March 2020].
29. Lefebvre, H. [1968] (1996) 'The Right to the City, *Writings on Cities*, Oxford, Blackwell, p. 147.
30. Lefebvre, 'The Right to the City,' p. 151.
31. Lefebvre, 'The Right to the City,' p. 152.
32. Lefebvre, 'The Right to the City,' p. 152.
33. See Shields, R. (1996) *Lefebvre, Love and Struggle*, London, Routledge, pp. 55–68.
34. Lefebvre, 'The Right to the City,' p. 154.
35. Gorz, A. (1980) *Farewell to the Working Class: An essay on Post-Industrial Socialism*, London, Pluto.
36. Lodziak, C. (2002) *The Myth of Consumerism*, London, Pluto.
37. Homebaked Anfield (2020) www.homebaked.org [accessed 6 March 2020].
38. O'Byrne, R (2019) quoted from *The Independent* [n.d.], www.home-baked.org.uk [accessed 29 February 2020].
39. Harrison, G. (2018) quoted in Pidd, H. 'Liverpool community laundrette honours the saint of the slums' *The Guardian*, 16 June, on-line [accessed 29 February 2020].
40. https://kittyslaundrette.org.uk [accessed 29 February 2020].
41. Lefebvre, 'The Right to the City,' p. 158.
42. Lefebvre, H. [1970] (2003) *The Urban Revolution*, Minneapolis (MN), University of Minnesota Press, p. 175.
43. Sandercock, L. (2006) 'Cosmopolitan Urbanism: a love song to our mongrel cities,' Binnie, J., Holloway, J., Millington, S. and Young, C., eds., *Cosmopolitan Urbanism*, London, Routledge, p. 48.

44. Personal memory (2011) Creative City Limits symposium, London, University College London Urban Laboratory.
45. Marcuse, P. (2009) 'From Justice Planning to Commons Planning,' Marcuse, P., Connoly, J., Novy, J., Olivio, I., Potter, C. and Steil, J., eds., *Searching for the Just City: Debates in Urban theory and Practice*, London, Routledge, p. 94.

INDEX

© The Author(s) 2021
M. Miles, *Paradoxical Urbanism*,
https://doi.org/10.1007/978-981-15-6341-6

Printed in the United States
by Baker & Taylor Publisher Services

Printed in the United States
by Baker & Taylor Publisher Services